MIGHTY STORMS OF
NEW ENGLAND

Oil on canvas painting of the Great Gale by John Russell Bartlett

MIGHTY STORMS OF NEW ENGLAND

*The Hurricanes, Tornadoes, Blizzards,
and Floods That Shaped the Region*

ERIC P. FISHER

Globe
Pequot
GUILFORD, CONNECTICUT

Globe Pequot

An imprint of Globe Pequot, the trade division of
The Rowman & Littlefield Publishing Group, Inc.
4501 Forbes Blvd., Ste. 200
Lanham, MD 20706
www.rowman.com

Distributed by NATIONAL BOOK NETWORK

British Library Cataloguing in Publication Information Available

Library of Congress Cataloging-in-Publication Data

Names: Fisher, Eric P., 1984– author.
Title: Mighty storms of New England : the hurricanes, tornadoes, blizzards,
    and floods that shaped the region / Eric P. Fisher.
Description: Guilford, Connecticut : Globe Pequot, [2021] | Includes
    bibliographical references and index.
Identifiers: LCCN 2021021635 (print) | LCCN 2021021634 (ebook) |
    ISBN 9781493043507 (cloth) | ISBN 9781493043514 (epub)
Subjects: LCSH: Storms—New England—History.
Classification: LCC QC941 .F57 2021 (ebook) | LCC QC941 (print) |
    DDC 363.34/920974—dc23
LC record available at https://lccn.loc.gov/2021021635

For my mom.
Thanks for buying me all those weather books as a kid.
It worked!

Contents

Preface

THERE IS JUST SOMETHING ABOUT THE WEATHER, ISN'T THERE? A MIL-lion things are going on in the world every day, but there are very few that we all experience collectively. Weather is one of them. From the most benign and beautiful day to the building crescendo of an unforgettable storm, there is usually something to talk about no matter where you find yourself on earth. You can stand in the freshening winds of a coastal gale or bake in the heat of a desert. Feel a thunderstorm breathe as warm air rises and cold air rushes down in response. Look closer and see the inexplicable detail of a snowflake on your sleeve or sit back to soak in a rainbow after a midsummer shower. Weather is endless (free!) entertainment.

Here is some great news for everyone: You do not have to be a meteorologist to appreciate weather or become an expert in it. All you need are your senses, a penchant for observation, and an occasional glance toward the sky. Humans have been using these skills for a very long time. Babylonians were attempting forecasts as early as 650 B.C. Roughly three hundred years later, a guy who had more than a few good ideas named Aristotle wrote *Meteorologica*—a collection of theories on the formation of weather phenomena. Quite a few turned out to be false, but we won't fault him for trying.

When it comes to forecasting though, a lot of what we know and use is startlingly recent. The first synoptic (large-scale) forecasting efforts did not begin until the late-1800s. The ability to communicate quickly and harness the power of observations to predict the future came decades later. Tornado forecasts did not begin until the late 1940s. After realizing, quite by accident, that radar picked up rain signals during World War II, NOAA rolled out the first set of radar dedicated to a national warning network in 1959. The first weather satellite named TIROS began orbiting in 1960, monitoring our planet from above

In other words, most of the tools we take for granted now are only a couple of generations old. And they're growing by leaps and bounds every day. Hours spent crafting a hand-drawn analysis of weather patterns has morphed into computer models rapidly digesting and pumping out heaps of data. Resolution of our satellite and radar imagery is allowing budding weather enthusiasts to watch storms unfold with unprecedented clarity. There is no doubt that it is an incredible time to be a fan of the sky.

That said, knowing the past is a critical part of understanding and forecasting the weather. Building a database of previous weather observations is called climatology, and it is a powerful tool. By amassing notes and records of storms and seasons, we can produce reasonable goalposts for any particular region on earth. Put another way, the data helps us know what to expect from the weather based on averages over time. This allows us to decide when to plant gardens, open up the pool, plan a leaf-peeping trip, or stock up on shovels and sidewalk salt. Perhaps more importantly, climatology tells us what the possible outliers are when it comes to extreme weather.

This is a book about outliers . . . specifically, outliers in New England. There have been a *lot* of storms in New England. The region breeds one of the highest concentrations of meteorologists in the country for a reason. One can experience just about anything except a dust storm. Snowstorms, floods, droughts, heat waves, arctic blasts, hurricanes, tornadoes, wildfires, and other atmospheric oddities come and go with the changing seasons. Rare is the boring year of weather. This is hardly an exhaustive list of every high-impact event. However, for one reason or another, these storms rise above the rest. It is a collection of high-water marks and snow piles. Ferocious winds and infamous bouts of destruction. They are all just a little snowier, wetter, windier, or hotter than the rest of the crowd.

By taking a walk down the road of high-end weather events, we can surmise what might lie ahead of us. Patterns in the atmosphere and oceans love to repeat themselves. Climate change is shifting the intensity and frequency of some weather events, but there always has been and will be extreme weather. A better approximation of "the ceiling" for potential natural hardships ahead helps us all better prepare. Plus, it is just fun to get your hands dirty in a stormy history!

Arguably most important of all, nearly all of these events left such an impression that they immediately led to progress. New warning systems have been implemented, government agencies formed, and technology accelerated in response to the damage and loss these storms left behind. Greater understanding of risk led to flood controls that were developed to save lives and property. Even something as simple as a regional group of mutual aid to prevent wildfires was born of one particularly extreme inferno. As destructive as many of the events were, they helped us move forward to protect society in new and innovative ways. For that, we can all expect to be better positioned for what will land on our doorstep next.

PART ONE
Winter in New England

Route 128, Boston, Massachusetts, Blizzard of '78

The first inklings that change is coming begin in August, when all the sudden we turn around and notice with alarm that the sky is getting dark a lot earlier. No other time of year seems to feature such rapid transition as when a brief fall descends into the dark and cold days by November. It is time for arguably the longest and most action-packed season of the year in New England: winter.

As the sun dips lower and its rays grow weaker, the meteorological magic begins in earnest. Cold air brews in the arctic and across Canada, sending advance scouts to fan out across the lower latitudes and chill winter's advance. By the time the solstice arrives in December, New England has usually seen snow several times and occasionally a memorable snowstorm decides to visit before Christmas. The months ahead bring all manner of precipitation, temperature swings, and storm types. If you want to be a meteorologist here, winter is where the sausage is made.

The trademark storm of the season is a nor'easter, but what exactly *is* a nor'easter? Many are surprised to hear it does not really have a strict definition. People cannot even agree that they're called nor'easters as the purists raise a fist to the clouds and yell "northeaster!" These storms are a mingling of ingredients that often come together just in time to pummel New England with snow, ice, rain, and wind. The nor'easter moniker comes simply from the fact that they have northeast winds where the impact is greatest, and not the direction from which they arrive. If we had to set a couple of qualifications, my personal opinion is that these northeast winds should produce at least *some* damage to earn the distinction. For good measure, let us say they also need to whip up the surf to noteworthy heights, too.

Nor'easters breed in fertile grounds where temperature gradients are large. If an arctic blast meets up with the warm waters of the Gulf Stream and the jet stream winds are just right, then in short order you will have your storm. They are not inherently wintry, but you are most likely to encounter them in the cold season from October to April. Most of the famous snowstorms you will read about fall into this category, and the fiercest ones undergo a process on every weather enthusiast's bingo card: bombogenesis. Not just a silly phrase from your local TV meteorologist, bombogenesis refers to a rapid drop in pressure. When a mid-latitude cyclone features a pressure drop of at least twenty-four millibars in twenty-four hours, it is official. You may also know these types of storms as "bomb cyclones" and they often bring high winds and storm surge. Get yourself at least three consecutive hours with winds over thirty-five miles per hour and visibility under a quarter of a mile and you have yourself a true blizzard.

The only thing that compares to a hurricane in terms of coastal hazards is a nor'easter, and the list of high-water marks on the eastern coastline of New England is primarily made up of cold season storms. Driving winds from the Gulf of Maine build massive swells, often leading to spectacular and destructive scenes when waves meet land.

Though these storms garner the most attention, there are all sorts of interesting aspects of winter besides the ability to wear sweatpants more often. The atmosphere is something that generally gets colder the higher up you go, but that is not always true. When a sliver of warmer air tries to invade, it takes the path of least resistance over the top of cold air and reverses the normal setup. Picture an unshaken jar of oil and vinegar salad dressing for this visual. Any time this scenario emerges during a storm, you can end up with sleet or freezing rain instead of snow, no matter what the temperature is at the ground. Sleet pings off the ground when it is cold enough below that slice of mild air for raindrops to refreeze into a pellet. Freezing rain occurs when the subfreezing air is shallow and near the surface, allowing supercooled water droplets to freeze on contact. Few sights are as breathtaking as the aftermath of an ice storm, but they are spectacularly destructive.

While we're talking about different things falling from the sky, did you know how many different types of snowflakes there are? Glad you asked . . . there are at least thirty-five different ways in which an ice crystal can form. From the classic "dendrite," which has a star-like shape and appears in holiday decorations, to the less endearing needles, columns, and plates. It all depends on the temperature and humidity of the environment in which they form.

Aside from the storms, it is simply a very cold time of year. The downwind side of a continent is a good place to tap arctic outbreaks, as it minimizes any path over open water to moderate air masses. The coldest temperature ever recorded in New England is -50°F, noted in both Bloomfield, Vermont, in 1933 and along the Big Black River of northern Maine in 2009.

That said, no season in New England is as variable as winter. Sometimes, essentially nothing happens all season long. There have been years when just several inches of snow fall. One year there may be several blockbuster storms, and others not a single nor'easter lumbers up the coast. The flakes can stack up over the course of months or arrive in a quick three- to four-week barrage. And occasionally, it is so warm that plants bud out in the heart of what should be their dormancy. Perhaps this inconsistency is what makes it arguably the most fascinating stretch of the calendar.

The Blizzard of '78

"Route 128 is unbelievable; it is as if there has been a massive rush-hour traffic jam and somebody had said, 'Stop' and covered the cars with 5 feet of snow."
—MASSACHUSETTS GOV. MICHAEL DUKAKIS, FEB. 8, 1978

EVERYONE HAS THAT "WHERE WERE YOU WHEN?" MOMENT OF THEIR lives. The moon landing, the Kennedy assassination, the day the Berlin wall came down, September 11. Seminal events burned into our memories. If there was ever a New England weather event that could elicit such a response, it is the Blizzard of '78.

A king sitting on a throne above all other winter contenders for Massachusetts and Rhode Island, this one snowstorm lives on in New England lore. If you have ever run out to get bread and milk before the snow starts falling, the Blizzard of '78 is the reason why. If the roads are shut down before a nor'easter advances, the Blizzard of '78 is why. Its effects are burned into long memories and likely will remain for many years to come.

What gives this one storm such standing? It was not just the snow, which was prodigious. Nor the waves, which were crushing. It was the cumulative effects of the weather, iconic images, and the advancing state of TV news with widespread coverage beamed into homes. It was one of the first examples of extended storm coverage in Boston; something that has become commonplace now. When all the ingredients came together it was a sight to behold, and, to remember for the ages.

Where do we begin? For starters, 1978 was a heck of a winter and it started before this infamous storm. You may search for the "Blizzard of 78" online and not even get the historic one that pounded New England, for there was another major snowstorm two weeks earlier in the Midwest.

From January 24 to 27 a colossal blizzard tracked into the Great Lakes, undergoing rapid cyclogenesis and shattering pressure records. It would bottom out at 955.5 millibars in southern Ontario, and the inland bombogenesis conjured up forty- to sixty-mile-per-hour winds with gusts as high as eighty-two miles per hour. The intense pressure drop earned the storm the nickname "The Cleveland Superbomb" while snowfall on the order of one to three feet buried Ohio, Indiana, and Michigan.

We can go back just a bit further and find yet more winter mischief in New England before the main event rolled into town. On the heels of a cold December, January was busy from the get-go. In the area of Hartford, Connecticut, over twenty inches of snow had fallen through the 18th with several consecutive storms swinging through. The accumulation of moisture-laden snow and ice produced near-disaster.

On the evening of January 17 nearly five thousand people gathered for a University of Connecticut vs. University of Massachusetts basketball game inside the three-years-young Hartford Civic Center coliseum. Just six hours after a UConn win, at 4:19 a.m., those in the newly built convention complex would describe a long rumble like an earthquake as the large flat roof gave way. Crashing down upon ten thousand seats below was a mess of steel, insulation, and ice. In a miraculous stroke of good timing, only two guards and two maintenance men were inside at the time and none of them were hurt.

The civic center was not alone in such a fate. The heavy snowfall brought down the roof of the Wire Wind Corporation in Jewett City, a supermarket in Windham, and a department store in Manchester.

And yet, there was plenty more ahead. Just two days later a major snowfall arrived, blanketing New England in one to two feet of snow. Often forgotten because of what followed, the snowstorm of January 20 and 21 was, at the time, Boston's second largest snowstorm since at least the late 1800s with 21.4 inches of snow. Only a February blizzard in 1969 had produced more in the city. Amazing to think that such an event could be left in the dust! But while the snow was significant and snarled travel, the winds and tides were not nearly as ferocious as the storm lurking two weeks out. Therefore, the impact would quickly take a backseat in the history books.

This mid-January snowstorm would also play an important role in the debacle to come, because it was a forecasting nightmare. Initially, the storm wasn't even expected to be a snow event but instead a rainstorm across eastern New England. The changeover to rain did not happen in some towns or happened after piles of snow had accumulated in others. There had been a couple other warnings of snowstorms that season that had not lived up to their billing. Seeds of doubt were put into the public's mind and recency bias can be extremely powerful. Though the Blizzard of '78 would feature much better forecasts, the perception was that it was overblown and would not live up to the hype. In the end, this is probably what led to the disaster that unfolded on roadways as the snowflakes quickly piled up.

Sunday, February 5, is where we pick up the story. A weak Alberta Clipper had been moving through the Great Lakes and toward the Mid-Atlantic during the day, eventually spawning a budding storm off the North Carolina coast during the evening. Forecasters had foreseen this development, using observations and the remarkably accurate (for this storm) LFM model, or limited fine-mesh model. For several days it had been printing out some startling data for what lay ahead. Though humbled by the failure two weeks earlier, meteorologists were quick to sound the alarm. The National Weather Service office in Boston put out Winter Storm Watches on Sunday morning, more than a day before the first flakes fell. The 5 a.m. issuance from the Boston office called for developing snow Sunday night and increasing intensity to follow Monday with considerable blowing and drifting. It closed with the ominous line "a substantial snow may come of it."

Blizzard conditions were mentioned fifteen hours in advance of the storm, with heavy snow warnings posted Monday morning. It may sound like short notice now, but this was a pretty good advance alert in the 1970s. The Monday morning edition of the *Springfield Daily News* asked, "Ready for 16 inches?"

The damage, however, had been done over the past several weeks. Confidence in the forecast was low and, it being a Monday morning, thousands headed out onto the roadways for their typical start to the week. Many of them would not take their car back home, but instead would abandon them on the side of whatever street they were snowed in on.

On this fateful morning, a strong arctic high pressure of 1052 millibars was drifting down out of Manitoba, anchoring in a cold air mass on the western side of a rapidly developing nor'easter. Perhaps surprisingly to anyone who is familiar with strong coastal storms off of New England, the drop was "only" down to about 984 millibars. There have certainly been much deeper storms. So why the ensuing ferocity? Considering the very high-pressure environment and arctic cold, it did not need extreme bombogenesis on the order of some of the biggest historical drops. The gradient, or change in pressure over distance, was still enormous and according to the work in Paul Kocin and Louis Uccellini's *Northeast Snowstorms* represented the largest gradient in their study of numerous snowstorms from 1950 to 2003.

At the upper levels, textbook transformation was underway. A polar disturbance dropped straight down from north of Hudson Bay on the 5th, digging down into the Great Lakes and then closing off at five hundred millibars, placing southern New England squarely in the northwest quadrant. For snow enthusiasts, this is the prime location for heavy snowfall.

Meanwhile the upper-level jet stream was undergoing a wholesale change. Leading up to the blizzard, a zonal west-east look had been in charge. Between February 5 and 6, a deep amplification took place as a ridge rocketed northward through the Plains and into Canada while a significant trough dug into the east. The analysis on the evening of February 6 revealed a coupled jet signature, which helps to force air upwards vigorously and aids in heavy precipitation. Everything was ripe for a major event from the surface on up.

As the roads filled with commuters and school buses, there was a final twist that would seal the storm's reputation. The first flakes began just a little later than forecast. Had they started to flutter down a few hours earlier, perhaps many would have had the chance to change their mind and stay home. But when the snow arrived around eight in the morning the day was underway, and the storm did not take long to show its hand.

Rates of two inches per hour were the rule by midday, and the full scope of what was happening became clear. Businesses and schools tried to get everyone out and back home as soon as possible, but the early dismissals were greeted by impassable roads and treacherous conditions. It was too late. Thousands of cars and their inhabitants were immediately caught on interstates and side roads along the Boston to Providence corridor. It would become one of the enduring scenes of the Blizzard of '78 as three thousand cars and five hundred trucks were trapped in the drifting snow along an eight-mile stretch of Route 128 alone. Another 1,950 were stuck on Routes 95, 195, and 146 in greater Providence. Many of the cars would stay in their snowbound state for days before help arrived to dig them out. While most people abandoned them in the snowbanks, some who decided to stay put were in a deadly situation. With the snow piling up and blocking tailpipes, fourteen people were killed by carbon monoxide poisoning.

The roadways were perhaps the most dangerous spot to be, but another location in downtown Boston was turning into a rowdy storm shelter as the day wore on. Throwing aside the forecasts, 11,666 fans had decided to tempt fate and have a few pints inside the Boston Garden. It was Beanpot time, the annual college ice hockey tournament for local bragging rights. Harvard took down Northeastern during game one, and during the action word was spreading about how wild it was getting outside. While it was starting to dawn on everyone that a major storm was taking place, more than a few decided to stay for game two between Boston University and Boston College. During that matchup an announcement was made indicating that conditions were continuing to deteriorate, but many still decided to stay put.

Toward the end of BU's thrashing of the Eagles, a new announcement took on a decidedly different tone. The message was to stay, because there was no getting out. All those cars were abandoned on the highway, the train system (MBTA) was shutting down, and it was going to be a slumber party at the Garden. Hundreds became storm refugees, and as it turned out it would not just be an overnight affair. The mix of leftover fans, Garden employees, and sports reporters slept in skyboxes and locker rooms for several days before they were able to make an escape.

A concession worker named Rich Fahey would later recall that it really wasn't that bad . . . in fact maybe it was a dream come true. Fahey wrote, "The temptation to stay put in the Garden was strong. There was free coffee, leftover hot dogs, and popcorn. We knew where there was beer to be had. Card games had broken out all around the club. Really, what else does a man need?" Fahey would also note that there were very few women since they had the common sense to leave or not attend a game during a blizzard in the first place, and that the stragglers were "pretty ripe" by the time they were kicked out three days later.

Some did make a run for it and ended up making some harrowing drives through blinding snowfall. By 11:30 p.m. snow depths around one foot were reported across eastern New England and the winds were howling. Through the night the blizzard raged, burying towns and thrashing the coastline with destructive surf and surge. It was during this time that the peak wind gusts tore across the landscape, reaching 92 miles per hour in Chatham, 90 miles per hour on Nantucket, 84 miles per hour at Blue Hill Observatory, 79 miles per hour in Boston, and an unofficial gust of 111 miles per hour in Scituate.

The full-throated northeast gale was directing the ocean's fury on the shoreline from Maine to Massachusetts, and Monday evening's high tide was the first test. Tides were running two to four feet above normal. It was a battering, but the worst was yet to come. The center of the storm had taken up a position just off the Long Island coast and had essentially stalled, performing a short loop and finding a spot due south of Narragansett Bay by Tuesday morning. While keeping bands of heavy snow flying and winds whipping, this slow movement stopped the Monday-evening high tide from going out in any significant manner. It set the stage for epic destruction as the sun rose for the next high tide, which was expected to be a foot and a half higher than the previous night.

A major variable for coastal flooding involves the moon. If a new moon phase is occurring, astronomical tides will be at a high point for the month (also known as a spring tide). Many meteorologists circle these tides on their winter calendar in advance to be wary of the threat. On February 6 not only was there a new moon but it was what we would now call a "supermoon," or a moon at perigee. A perigean moon is at its closest point to earth in orbit, which adds an extra gravitational boost to the already high astronomical tides. In short, the timing could not possibly have been any worse.

With a blocking high slowing down the storm and piling up water levels over multiple tide cycles, the ocean's refusal to recede at low tide becomes a dangerous risk. The biggest flood events from nor'easters tend to be ones that occur over several cycles. Having both the blocking and new moon in tandem is a recipe for disaster.

An early-morning NWS bulletin issued at 5:30 a.m. summed up the grim situation for the coastline, warning that "considerable and extensive flooding is probable later this morning" and "extensive surf battery and beach erosion will continue throughout the day." It would indeed be unlike any storm in memory.

From Revere just north of Boston, to the South Shore and Cape Cod, all-time high-water marks pushed over flood barriers and destroyed property. The coastline was reshaped by the powerful swells breaking through barrier beaches and rushing through weak spots in the dunes.

Winthrop Drive in the Beachmont section of Revere appeared as an icy lake when the Atlantic came over the seawall and submerged cars in a slushy mix of saltwater and snow. Homes bobbed in the floodwater that had leapt over the inadequate seawall, and three thousand people would need to move into the Revere High School when it was opened as a storm shelter. The towns of Scituate and Hull, facing directly into the full brunt of the waves, were devastated. Homes were smashed and splintered, thrown off their foundations. From the Boston area southward, records fell.

Seaside homes in the Gun Rock Beach section of Hull were torn apart in an unimaginable scene. Down the road in Cohasset, the Old Ice House near Hugo's Lighthouse restaurant had been documenting high-water marks since 1851. None marked on a board in the icehouse were as high as the morning of February 7, 1978. Boston's tide set a new record that stood for forty years until January of 2018. A refurbished 269-foot Hudson River cruise ship, the *SS Peter Stuyvesant*, sank next to Anthony's Pier 4 restaurant along Boston Harbor. It would remain in its watery grave until 2017 when the leftover remains were finally brought up from the depths during waterfront construction.

A town known for being ravaged by coastal storms, Scituate, featured some of the most dramatic images. Streets were full of boulders and debris and the wreckage of homes was strewn all over shoreline roads. One photo published in the *Boston Globe* featured a house with only one fixture left remaining—the toilet.

The solid concrete seawall in North Scituate was broken apart as if it were made of sticks, and the Peggotty Beach section of town was left in ruins. In all, an astounding two thousand homes were destroyed by the savage surf that Tuesday morning. The losses included two iconic structures, one on Cape Cod and one on Cape Ann. The Fo'castle, or "Outermost House," immortalized in naturalist Henry Beston's 1928 book by the same name, was tossed into the ocean from its dune in Eastham. To the north, it was Motif No. 1 that was swallowed by an indifferent sea. Known by artists as "the most painted building in America," the alluring red fishing shack had stood for 138 years (an exact replica was rebuilt later in 1978).

Sad as the loss of a fixture in town was, Rockport and the rest of Cape Ann had other more pressing problems. Cars were flung into the harbor and houses in Motif No. 1's neighborhood of Bearskin Neck were crushed.

To the south on Cape Cod, the morning brought an eerie calm as the center drifted just off the coast of Nantucket. The nor'easter had taken on an appearance similar to a hurricane, with a clearing "eye" in the center visible even on the more rudimentary satellite imagery of the 1970s. With breaks of sunshine, the damage Cape Codders were waking up to was immense. Whole portions of the coast had been rearranged by the blizzard.

At Coast Guard Beach in Eastham, the Outermost House was not the sole loss. Waves obliterated the dunes and destroyed the bathhouse as well as the parking lot for the beach. The overwash from storm surge blasted into Nauset Marsh, depositing the remains of several "camps," or simple beach shacks, behind the old dune line.

In Chatham, Monomoy Island was split in two in another case of the storm altering New England's coastline. The newly separated North and South Monomoy Islands did not join again until November of 2006. A jog up the coast to North Beach in Chatham was a scene of devastation with several other camps and cottages destroyed overnight.

Another historic split took place in Truro, farther north along the National Seashore. At Ballston Beach, the ocean surged through a low point in the dunes and separated Pamet Road. The town decided not to reconnect them, and so North and South Pamet Roads are still without a link today. The next and final town up the coast is Provincetown, where major flooding inundated Herring Cove Beach and Commercial Street in the downtown area. The Outer Cape was not alone as major flooding also inundated towns on the Massachusetts Bay side, particularly in the vulnerable areas of Sandwich and Dennis.

In all, flooding occurred over four consecutive high tide cycles with the peak destruction coming Tuesday morning. Few storms over the history of New England settlement rivaled the damage these successive high tides wrought on the coast.

Those inland were not enduring the battering ram of an angry ocean, but they were watching an otherworldly amount of snow pile up. A strong blocking high pressure over eastern Canada and Greenland was working its magic in two different ways. Cold arctic highs can keep a fresh drain of frigid air pouring southward into the storm, keeping snow as the favored precipitation type. Blocking also slows down a storm, increasing the longevity of poor weather conditions.

The duration varied from town to town, but in general the snow fell continuously for thirty-two to thirty-seven hours before the storm finally slid away to the east and came to an end Tuesday evening. The final flake was noted to fall just before the stroke of midnight at Blue Hill Observatory. What was left behind would become a story to tell for generations.

In Boston, it became the biggest snowstorm on record (at the time) with 27.1 inches choking city streets and essentially shutting down daily life. It was the first time mail was not delivered in the area since the great Hurricane of '38. Plows were completely overwhelmed and there was no place to put all the snow anyway. Governor Dukakis shut down all travel except for emergency personnel for several days, a ban that would not occur again until a blizzard in February of 2013.

A swath of two to four feet of snow had accumulated where banding had been persistent Monday night, from Cape Ann (32.5 inches in Rockport) through the western suburbs of Boston and south to Rhode Island and northeast Connecticut. The capital city of Providence had an eye-popping and record-setting 28.6 inches of snow, the most intense twenty-four-hour snowfall on record. But the real jackpot was just north. The official measurement for Woonsocket, Rhode Island, was thirty-eight inches, though unofficial reports went as high as fifty-five inches in the town of Lincoln. Wind had whipped the snow into towering drifts fifteen to twenty-seven feet high. While not in the heart of these incredible snow totals, a broad region of one- to two-foot snowfall was noted across Maryland, Pennsylvania, New York, and the rest of New England. While still significant in Connecticut, the blizzard came up short of the Great White Hurricane of 1888, which arguably remains the most severe snowstorm of record west of the Rhode Island border.

Without a doubt, this was great fun for many children and those young at heart. For some of them, school would not be back in session until the middle of the month given the magnitude of cleanup required. Memories of deep snow drifts you could jump off the roof into and memorably enormous snowmen are common.

There was also a danger with so much snow on the ground. One of the saddest stories to emerge came from the small town of Uxbridge, Massachusetts. A ten-year-old boy went missing on February 7 and the tragic truth of his location was not uncovered until the snow melted. Even with hundreds of volunteers digging through and moving around snow to search for the young boy, no sign was found for three weeks. It was then that mailman Leo Lussier caught a glimpse of a mitten sticking out of a snowbank while on his route. It was Peter Gosselin, found just feet from his parent's door. The death toll was sobering for a winter storm, with seventy-three killed in Massachusetts and twenty-six more in Rhode Island.

This was a job that called for much more than the shovel or snowblower. National Guard troops from Fort Bragg and Fort Devens were called in to help with the effort and clear Logan Airport's runways. Even the Army and Air Force moved in from southern bases to add their manpower and equipment to the task. President Jimmy Carter declared portions of Massachusetts, Rhode Island, and Connecticut federal disaster areas. At best, it was days before travel started moving again. At worst, it took over a week before the roads were cleared. Shelters were brimming with over ten thousand temporary residents waiting for heat and electricity to return, or to find a more permanent option if their home had been destroyed. The damage left behind totaled well over $1 billion in today's dollars.

The tales of scores of New Englanders trudging through waist-deep snow, sleeping in offices or cars, and pitching in to reopen the region are too numerous to recount. Though there is no arguing it affected the psyche of everyone in its paths for decades to come. Nature did, at least, give everyone a break after the incredible stretch of severe winter conditions. It did not snow again in most towns for the rest of the month. Instead of the calm before the storm, it was a calm after the storm.

Vehicles stranded in snow in the southbound lanes of Massachusetts Route 128

The Great White Hurricane of 1888

"Chaos reigned, and the proud, boastful metropolis was reduced to the condition of a primitive settlement."
—THE NEW YORK TRIBUNE, MARCH 13, 1888

GO BACK OVER THE PAST SEVERAL HUNDRED YEARS AND YOU WILL FIND no shortage of memorable snowstorms across the US. Like clockwork the arctic fades into darkness, frigid air masses descend, and snowflakes fly. Among the innumerable storms that have left their mark on our lives and property, two of the worst hit in the very same year. The first would lead to tragedy in the heartland, with the follow-up striking the east coast two months later. That year was 1888.

It began in early January, when a bitter air mass engulfed the Dakotas and Minnesota with temperatures plunging -20 to -40°F. Brutal as it was, by morning of January 12, the air had moderated, and it was easy for many to let their guard down when venturing out in the more tolerable conditions. Many accounts speak of the pleasant and calm start to the day without any hint of trouble. This would prove to be a fatal error in judgment as an arctic front was on the move from the north. Rapidly, a whipping blizzard set in immediately trapping thousands of people. It was in the schoolhouses, however, that the most gut-wrenching part of the storm played out. Confronted with the question of whether to stay and risk running out of food and heat or try to get back home, teachers and children became victims. It is estimated that 235 died in the sudden storm, a large number of them students who were overcome by the blinding snow and cold while making a run for it. This toll earned 1888's first major storm the title of "The Children's Blizzard."

After this sad start to the winter season, attention would turn to an even more deadly storm destined to strike the east coast in March. At the time, it did not feel like winter's wrath was coming back for another round. In fact, the setup was very similar to the Plains blizzard in January. A cold shot that kicked off the month of March had relaxed and moved on, allowing for a feel of spring to creep into the minds of New Englanders and New Yorkers. It was the start of the weekend and an historic snowstorm of unrivaled proportions was not on the mental menu of what lay ahead. Saturday, March 10, featured the optimistic signs of crocus popping up, grass greening, and birds singing. Temperatures popped into the forties and fifties. Everyone was ready to move on.

There had not yet been the creation of the Weather Bureau or Weather Service, but a predecessor to these institutions was monitoring the atmosphere. In 1888 the Signal Service, a branch of the War Department, did have a contingent of fledgling meteorologists attempting to gather observations and make daily forecasts. Using the tools available at the time, it was noted that a trough of "low barometer" extended from southern Canada to the Gulf of Mexico with two more distinct areas of low pressure on the surface analysis—one in the Great Lakes and the other over Georgia. For a current meteorologist, this would probably get the senses tingling. Right away one could envision a merging of these two low centers into one large coastal storm. The thought did not occur to forecasters at the time. The outlook they issued on the morning of Sunday, March 11, was, in a major understatement, not good.

"Fresh to brisk easterly winds, with rain, will prevail tonight followed on Monday by colder, brisk westerly winds and fair weather throughout the Atlantic States."

Indeed, it did begin with brisk winds and rain in New York City. On Sunday, the rain began to fall and when many city dwellers went to bed there was nothing much to worry about. The scene upon waking, on the other hand, was a completely different matter. An opening of the blinds revealed a raging whiteout. Nearly a foot had already added up and the drifts were rapidly climbing higher.

Over the next two days, an epic snowstorm would smash records, tear down telegraph lines, cripple public transportation, kill more than four hundred people, destroy two hundred ships at sea, and cut towns off from one another. Many of these towns have never come close to a similar amount of snow from a storm again. A call for rain instead became one of the greatest blizzards ever witnessed in the east.

So, what happened? Like so many of the major events chronicled in this book, the storm slowed down, stalled, and looped off the coast to bring nearly three full days of high-impact weather. We see the same story play out with the Blizzard of '78, the Perfect Storm, and more. Such blocked and sluggish movement is what allows storms to dump extremely unusual amounts of moisture and whip waves up into a destructive force of coastal surge. The 1888 blizzard was no different than other extreme examples over the course of time, but outcomes were at the top of the scale when it comes to what is possible in the region.

By the evening of Sunday, March 11, coastal reorganization was underway east of the Outer Banks. The weak surface low across the southeast saw its pressure fall from 1010 millibars in the morning to 1004 millibars in the evening. At 10 p.m., the ship *Andes* recorded a minimum pressure of 994 millibars just east of the Outer Banks, translating to a sixteen-millibar drop in approximately twelve hours. The sudden drop intensified winds along the Mid-Atlantic coast with the Signal Service reporting significant structural damage including roofs blown off and trees uprooted. Ships at sea were holding on for their lives but also providing the crucial service of continuing to observe the deteriorating conditions. The ship *Kensett* noted a pressure of 993 millibars at 7 a.m. Monday morning just east of the Jersey shore, with yet another vessel east of Long Island finding a minimum pressure of 984 millibars by 3 p.m. in the afternoon. Rapid cyclogenesis, in this case bombogenesis, turned a trough of low pressure into a potent nor'easter overnight.

The great squeeze was on with the deepening storm and an arctic high of 1040 millibars building in from the Great Lakes, the increasing gradient building strong northwesterly winds and dragging arctic air down from the north. The transformation was swift. Chilly rain in New York turned to snow by 1 a.m., and New Haven, Connecticut, by 2:30 a.m. A developing and extreme coastal front moved in from the ocean and set up across eastern New England, oriented north-south. In this case, the change in temperature on either side of the front was amazing as it separated the mild ocean air from dry and frigid arctic air. At one point, the range between Nashua, New Hampshire, and Northfield, Vermont, was 34° to 4°F. On a more localized basis, surface temperatures quickly dropped 10°F to 20°F west of the coastal boundary.

These coastal fronts act as a focus for moist Atlantic air, a convergence that sends air parcels up and precipitation down. Once meeting the cold air west of this convergent zone, the snowflakes came in droves. Throughout "Blizzard Monday" it refused to budge and prolonged whiteout conditions enveloped western New England to New York and New Jersey.

While the snow and winds raged, the low itself had stalled and was drifting aimlessly off the south coast of New England. Moving to a position just south of Martha's Vineyard during the afternoon of the 12th, the storm began to drift back to the west along the Rhode Island coast where minimum pressure bottomed out at an estimated 978 millibars. During the 13th the center slid southward, now to a location just south-southeast of Montauk, New York. Snow was still flying but became disorganized with intensity gradually diminishing. It wasn't until the 14th that the storm began to dissipate and slide slowly away to the east. We do not have reliable upper-air maps of the time, but it can be assumed that there was significant North Atlantic blocking to account for such a slow, retrograding motion.

The stall and loop observed during the 12th to 14th was a critical element to produce outrageous amounts of precipitation, and the resulting totals were nothing short of extraordinary on the ground. The very phrase "30 to 50 inches of snow were common" is not something one will often hear anywhere in the eastern US, but it was the real deal once the tumult of the storm relented. The Hudson Valley would end up being a bull's-eye.

It had set in on Sunday afternoon, and when all had ground to a halt Saratoga Springs, New York, was at the top of the list where up to fifty-eight inches of snow was reported! In nearby Troy, fifty-five inches fell and Albany observed 46.7 inches. With records dating to 1884 in Albany, no other storm has come within twenty inches of the Great White Hurricane. It made for the snowiest March on record and second snowiest of any month in the city, just behind December of 1969. Everything was closed as the Capital Region went into total shutdown.

In New England, it was a split between two types of snow. The milder side of the coastal front featuring heavy, wet snow that ripped down telegraph lines and required much hard labor to clean up. Boston oscillated between rain and snow, in the end only receiving seven to twelve inches across the city depending on proximity to the water. Even spared the worst thanks to a mixed precipitation, railroad lines were covered with debris and shut down. Communication by way of telephone or telegraph was taken out as lines fell. The paste of wet snow easily made quick work of trees and infrastructure. The *Boston Globe* ran the headline "Cut Off" while proclaiming it to be the most severe storm in a generation. Farther to the southeast, the whole event was one big windy rainstorm. A quick jaunt inland on the colder side of the storm was a completely different matter.

Just a dozen miles or so away from the coast, the snow totals quickly climbed above a foot. Upon entering the Worcester Hills, they approached three feet. The city of Worcester itself was buried under thirty-two inches of new snow and drifts more than double that depth. Keene, New Hampshire, notched thirty-eight inches. Respectably huge as these amounts were, the piles got bigger and bigger moving into the frigid side of the storm across western New England where temperatures barely held above 0°F for much of the event. The combination of arctic air, winds gusting thirty to fifty miles per hour inland and as high as eighty miles per hour near the coast, plus historic snowfall made for severe and unrivaled blizzard conditions.

The Connecticut River Valley is a region often "skipped" by big snowfalls that are more frequent in the higher terrain on either side. In 1888, this was not the case. Throughout the Massachusetts portion of the valley, thirty- to forty-inch totals were common with drifts well over six feet. A Springfield man named Edward Leonard happened to reach into one while trying to recover his hat and found instead a girl buried underneath the snow. He frantically dug her out and brought her to safety, where she did manage to survive thanks to quick action. A famous photo in Northampton captured pedestrians walking through tunnels in the snow on the campus of Smith College with a solid six feet of headroom accommodating most adults. Telegraph boys working in the storm tied wires around their waists so that they could be pulled out of the drifts.

Amazingly, the thirty to forty inches found here were on the low end when comparing towns just to the north or south. Southern Vermont was crushed with up to four feet of snow, including forty-eight inches in Bennington. A train on the Boston and Albany Railroad became trapped along the Vermont and New Hampshire border. The seventy-two passengers on it needed to be resourceful, because they would not be dug out of the towering drifts for two days. Their survival diet included eating raw eggs that were found inside a crate in the baggage car. Meanwhile a train stuck in Pittsfield, Massachusetts, lost thirty-two cars full of pigs that froze to death in the subzero temperatures.

Closer to the center of the storm in Connecticut, biblical amounts of snow piled high. A staggering fifty inches was reported in Wallingford, 44.7 inches in New Haven, and 42 inches in Middletown. To reach this unprecedented amount of snow, the blizzard was able to conjure up over five inches of liquid equivalent (meaning if it were rain, it would have been five inches of rainfall). The cold atmosphere of the winter rarely produces such significant moisture and it was only possible by way of the storm's slow movement and extreme banding along the coastal front. In the hardest hit of these Connecticut and Hudson Valley locations, the drifts were reported to reach a towering height of thirty to forty feet, including one measured to be thirty-eight feet high in Cheshire, Connecticut.

Of course, this meant no one was going anywhere in a hurry and many were plunged into survival mode. One story from Connecticut tells of a Mrs. M. Brusselars who could not escape her home in Hartford. Stuck for three days with a collection of twelve other storm refugees, the group resorted to eating local birds for sustenance. She claimed, "we found that under my back porch about seventy-five to one hundred sparrows had gathered, so we killed some of them, made a few sparrow pies, which helped to sustain us."

On a farm in Redding, so transformed was the landscape that "snow was so deep in the stables that cows and horses stood to their middles in it." Two-legged folk took to ladders to climb down out of their second-story windows to reach the ground below, since the first floors were choked with drifting snow. Everything from mail and milk deliveries to railroad service came to a stop and a mass effort toward snow removal became the primary objective. Until rail could return to normal, food shortages were an issue and got worse in the days to come.

New York City was a complete and total winter wasteland, isolated in its cramped and snowbound situation. In recent years there had been a rapid and haphazard buildup of telephone and telegraph wires that were strung in every which manner. The blizzard made short work of them, sending lines into the streets and providing an extra blockade in addition to the snow. Many fire stations were immobilized as a result and the property loss from fires alone tallied $25 million.

There were twenty-one inches reported in Central Park but some boroughs including Brooklyn and Queens reported up to thirty-six inches of snow, which can be much less manageable in tight city quarters than in the wide-open rural areas. Of the more than four hundred who died in the storm, half were in the city. Many of them were found buried in snow drifts along the sidewalks having been disoriented in the whiteout or giving in to hypothermia. One of the deceased was a New York senator, Roscoe Conkling, who could not survive the walk from Wall Street to Madison Square.

Samuel Clemens, more famously known as Mark Twain, was trapped in a Manhattan hotel for the storm. Always with a talent for describing the ups and downs of New England weather, he clearly was growing restless and annoyed. Instead of soaking in the greatest snowstorm in generations, he wrote to his spouse that he was "out of wife, out of children, out of line, and out of cigars." The humanity!

Nature was not inclined to lend a helping hand for the recovery effort and instead served up unseasonable cold for much of the rest of March. It is still the coldest March ever recorded in New York City's Central Park, dating back to 1869. April was not much help either as spring warmth was in very short supply. Almost equally as extreme, it still stands as the third coldest on record.

Without a doubt, the cold that funneled down from the north was an equal part of this epic blizzard. New York went from a balmy 42°F on Sunday afternoon (the 11th) to 6°F in the predawn hours of Tuesday morning. It remains the coldest temperature ever recorded in the city so late in the season. In Connecticut thermometers plunged subzero, as cold as -15°F, which resulted in the death of livestock and wild animals alike throughout the snow-filled frigid woods. It is likely that this level of cold produced a very powdery snow where totals topped three feet, allowing it to stack up and drift easily versus the wet and heavy snow that was more efficient in tearing down trees and infrastructure to the east.

Witnessing the annihilation of communication wires and the shut-down of public transportation, the Great White Hurricane of 1888 was a pivotal moment for the future of cities. Up to fifteen thousand had been trapped on elevated trains during the storm in New York. No one wanted to have a repeat performance whenever a big storm rolled into town again. In New York City and Boston, plans were hatched to put more utility lines underground and to look subsurface for travel needs. If recovery was going to happen more easily in the future, then heading below the soil and away from Mother Nature's punches was the way to go.

These plans were put into action with surprising expediency. The first subway in America would become reality just nine years later in Boston, now known as "The T." One of the most extensive in the world would begin shortly thereafter in New York City. Without needing to rely on horses, wagons, or trolleys, this new mode of transit enabled at least partial movement of workers and residents in big cities during the innumerable storms over the following decades.

A botched forecast was another opportunity for progress. In a post-storm silver lining, a lot more thought was given to the future of forecasting. Since the Signal Service had failed in its outlook, it was decided the time was right to depart the War Department and enter into a new realm where meteorologists could focus strictly on the weather and improve their prognostications. On October 1, 1890, President Benjamin Harrison signed legislation creating the Weather Bureau under the Department of Agriculture. There would be more failures ahead, but the road forward had been paved toward more reliable and lifesaving outlooks. The Weather Bureau would house the nation's best forecasters and held on to the name until adopting the name of "National Weather Service" nearly a century later in 1970.

Notable Out-of-Season Snowstorms

"Winter is begun here, now, I suppose. It blew part of the hair off the dog yesterday & got the rest this morning."
—Mark Twain—letter to Chatto and Windus,
October 21, 1892

We like to joke about the weather in New England, mostly because it tests our patience frequently. The old saying about the seasons is that we enjoy almost winter, winter, still winter and road construction.

Yes, winter is the signature season and can be a grind. It often snows on and off November through March and the warm sunny days of summer seem far more fleeting. Many people do not even bother taking the ice scraper out of the car.

Blockbuster storms like 1888 and 1978 tend to reside in the heart of winter, from mid-January to mid-March. But like hurricanes, peak season is not an exclusive window. Occasionally, the currents conspire and allow flakes to fly before trick-or-treaters head out or after the Red Sox start playing ball at Fenway Park. Here we look at two extreme snowstorms that caught even wary New Englanders off guard.

We start with the winter of 1996 to 1997, which was not the most exciting in memory. December arrived in mild fashion. After a seasonable January, an extremely mild February took over. At the time, it was the fifth warmest winter (December through February) on record in Providence, Rhode Island, and eighth warmest in Hartford, Connecticut. Heating and plowing bills were low.

Snowfall was below average; many locations with about two feet or less for the season. By the end of March the sun was getting stronger, crocus had bloomed, and the solstice had passed. Safe to say, a lot of people were probably left thinking they had gotten off pretty easily after a record-setting 1995 to 1996 winter and were sniffing spring around the corner. Hardware stores had begun putting away shovels and snowblowers in exchange for pansies and topsoil.

The weather, however, had other ideas. On the final eve of March, a storm that would break all the rules for the time of year was brewing.

The April Fools Blizzard, as it came to be known, was not classic in the textbook sense. There was no big cold high sitting over Canada pumping in the arctic air. Barely any cold around at all, in fact. Highs were in the fifties and sixties March 26 through 30 and still well into the forties on the 31st after a cold front came through. A false sense of security for both everyday folks going about their business and forecasters alike.

If this storm was going to produce tree-snapping snowfall, it was going to have to manufacture its own cold air. And that is precisely what it did.

The source of colder air was up across southeastern Canada. There it would linger, waiting for a call to action. At the same time the storm center itself began to take shape in Virginia and made a move to the coast, deepening off the Mid-Atlantic. Looking at radar around this time there would have been a lot of green showing. Light to moderate rain was spreading across Pennsylvania and New York, en route to New England.

By morning of the 31st, a different kind of system was blossoming. The pressure rapidly dropped off New Jersey, down to near 980 millibars. Rain and wind intensified, lashing the region. And as the raindrops fell, the cold began to take shape.

There are two ways in which a storm can "create" its own cold. Either by advection or dynamic cooling. Both processes occurred with the April Fools Blizzard to help offset a very mild start.

Advection essentially describes the movement of air, in this case from a source of cold toward an area of mild air. As the storm strengthened it began to tap the region of wintry chill lurking over southeastern Canada and drew it southward into New England.

Dynamic cooling occurs during heavy precipitation, especially when there is strong vertical motion. Think of all the little air parcels around you shooting upward into the sky. As they rise, they cool. And the faster they rise, the more efficiently they cool off. These regions of strong upward motion are usually found on the northwest side of our coastal storms, which is exactly where New York and New England were sitting relative to its center.

The other element of dynamic cooling comes from frozen particles like ice and snow melting. Oftentimes, the air thousands of feet above our heads where the clouds are is very cold and snowflakes form. But near the ground it is too warm, and they all melt before hitting us. You can imagine that temperatures near 50°F on March 31 were not hospitable for snowflakes to survive. But by melting there is energy being transferred. The process of changing a solid to a liquid takes heat out of the air, and so over time a net loss of energy cools the air column.

Once the warmer antecedent conditions had been altered, the snow came down in earnest. By the time everyone was finishing up dinner, areas away from the coastline had picked up about a half foot. As the night wore on temperatures continued to drop and the storm drew in huge amounts of Atlantic moisture. Thunder boomed and lightning lit up the sky with numerous reports of thundersnow across the region.

The rumbles of thunder began to mix with the sound of cracking and falling trees when the heavy wet snow gripped everything it landed on and added its weight. As the hour approached midnight snow was falling at a rate of an inch per hour, but it only got worse from there. Between just 11 p.m. on the 31st and 3 a.m. on the 1st of April, the snow came down at an absurd rate of three inches per hour. A full foot accumulated in Boston during just four hours overnight.

When the snow tapered off during the morning, 25.4 inches in all made the storm one of the biggest on record for the city, and it wasn't even winter! The twenty-four-hour snowfall record (for any time of year) still stands.

In central Massachusetts, where the rain had changed over to snow more quickly, the piles people woke up to were astounding. The city of Worcester notched its biggest snowstorm on record with a whopping 34.5 inches. It remained the largest snowfall there until a blizzard in January of 2015.

A little to the southeast, the jackpot was measured in Milford, Massachusetts. A full three feet lay on the ground, pasting trees and power lines. Burrillville, Rhode Island, came in at thirty-one inches, and in New Hampshire twenty-seven inches was reported in Jaffrey. It was no slouch for April in Connecticut but not quite as severe, with over twenty inches in the northeast hills.

The pure amount of water locked up in the snow was impressive. The slow-moving system was able to tap the Atlantic and dump unusual amounts of H_2O, both liquid and solid, across southern New England and New York. Snow is melted to find a "snow water equivalent" or SWE, since all snow is not created equal. For instance, lake-effect snow may have a ratio of 20:1 or 30:1, meaning twenty to thirty inches of snow per one inch of water. We also see this during very cold snowstorms. But for a sopping mess of a snowfall like on April 1, it was a much lower ratio. Total liquid equivalents of three to four inches to get two to three feet of snow were common, with Hull, Massachusetts, reporting 5.32 inches! Essentially, a full month's worth of moisture or more fell during the storm.

Very wet snow also comes with a different risk—heart attacks. This is where the phrase "heart attack snow" comes from and sadly it came to fruition here. There were three confirmed deaths from the storm across Massachusetts and Rhode Island, all due to overexertion while shoveling.

Strong winds helped to rip down trees and power lines pasted in the sloppy snow. Gusts were clocked at seventy miles per hour along the coastline, taking down part of a local and national landmark in Boston Harbor. *USS Constitution,* affectionately known as Old Ironsides, had just come out of a several year dry dock stint for restoration. The job was nearly complete with the final re-rigging underway when the storm hit. The wind was strong enough to snap the topgallant/royal mast. Fortunately, it was repaired in time for a two hundredth anniversary sail that summer.

With so much damage needing attention thanks to downed trees and power lines, Massachusetts Governor William F. Weld declared a state of emergency. Commuter trolleys were cancelled in Boston for the first time in two decades. Across the northeast seven hundred thousand customers lost power. Thousands were stranded at Logan Airport, which shut down for a time to remove snow from the runways.

The saving grace was the time of year. Considering it was not deep winter, the high sun angle of April and warmer temperatures made quick work of the snow and cleanup somewhat easier. Highs were in the fifties and sixties for the next several days and spiked into the seventies on April 7. That was that! But it remains a very memorable storm to this day and keeps DPW crews wary of putting the plows away until April.

THE AUTUMN EQUIVALENT...

When it comes to major winter storms arriving late, the April Fools Blizzard takes the cake. But how about a storm crashing the party early? For this, we look to the bizarre storm of October 2011, which was to become known as "Snowtober." It may very well have been the biggest October snowfall in the region since the Little Ice Age.

Perhaps it is more understandable to get a snowstorm during early spring. The atmosphere is coming out of its coldest state, the ocean is still very chilly, and our state of mind is still on alert from recent experience. We rarely just blast into warm weather during late March/early April anyway. But autumn? It is foliage trips and hayrides and enjoying the final warm days before settling in for winter. Shorts and T-shirts are still holding out hope in the dresser before getting tucked away for the season.

No one would be surprised by some early novelty flurries or snow showers to whip up some excitement, but a major snowstorm is still a distant expectation. Oddly, the storm that would drop flakes from Virginia to Maine played out like it was the middle of January and skipped right past this introductory phase.

On October 27, the table setter swooped through New England bringing a shot of cold air and even some accumulating snowfall, much like an arctic front would do in winter. Even still, it was October. It wasn't *that* cold. And that was the really surprising thing about "Snowtober." A lot of the accumulation took place without even getting to the freezing mark at the surface. Many locations that were stacking up snow and snapping trees hovered around 33 to 34°F for the entire event. The heavy rate of snowfall more than made up for it.

This classic nor'easter rapidly deepened off the Mid-Atlantic coast and tracked just offshore to the south of Nantucket, Massachusetts. All the while making the most of this just-cold-enough air.

Even though it was so out of character for the time of year, it must be noted that the storm itself was well forecast. Winter Storm Warnings were issued by the National Weather Service for the whole corridor of expected snowfall. Utility companies got ready with extra crews on call. Shelters were opened in the state of Connecticut.

But contrary to an early spring snowstorm, there is one glaring difference when it comes to impact. Leaves. Giant, leafy canopies ready and willing to accept the weight of snow. Up to this point the autumn had been so warm that many of the leaves weren't even sporting their vibrant yellows and reds yet. They were still green and healthy! And about to take a beating.

Much like forty-mile-per-hour wind will cause significantly more damage in summer than winter, an equal snowfall will do much more damage in fall than winter. The destructive power of "Snowtober" was unmatched in recent memory. National Weather Service forecaster Bill Simpson summed up the storm by telling the *Boston Globe* that "Fifteen thousand years ago, in the Ice Age, I'm sure they had more snow. But for modern day, this is unbelievable."

In New York City's Central Park, it was the first time on record that over an inch of snow accumulated during October. The 2.9 inches storm total doesn't look like much on paper but it proved to be one of the most destructive storms the park had ever witnessed. Heavy, wet snow destroyed a thousand trees. A warm autumn season meant that the trees were covered in (mostly still green) leaves that collected the weight of the snow. Uptown in the Bronx, an estimated 2,200 trees were damaged in the New York Botanical Garden.

The state of Connecticut had notched its largest power outage in history just two months earlier during Tropical Storm Irene. "Snowtober" was much worse, knocking out power to over 800,000 customers in Connecticut. The state's transportation department estimated that five times more trees were knocked down compared with Irene, which at the time had been a nightmare of damage and power outages. To surpass Irene, and so quickly, was a testament to its unprecedented nature.

The core of highest snow totals was found across Connecticut, Massachusetts, and New Hampshire. For many, the wet and destructive snow started to turn more powdery as the storm raged on, making for some flat-out ridiculous snowfall amounts for any time of year, let alone October. So far as reliable reports go, the king of the hill was in Peru, Massachusetts, a small town in the Berkshires. An incredible 32 inches fell there, with a similar amount found in Jaffrey, New Hampshire, at 31.4 inches.

Though these amounts are the highest, some of the worst impacts were found where the snow never turned powdery and the weight was far too much to bear for trees and power lines. Much of central Connecticut had just over a foot, but power outages in the wake of the storm would drag on for days with whole tracts of power infrastructure in ruins. The slow slog of power restoration would cost Connecticut Light & Power's CEO, Jeffrey Butler, his job. The final Connecticut Light & Power customers received power back on November 9, eleven days after the storm.

The sheer improbability of Connecticut enduring the two worst power outages in its history within two months of each other is remarkable. But while both Tropical Storm Irene and "Snowtober" left people in the dark for over a week and a half in many communities, there was one big difference between the two storms. One hit in late August, the other late October. The conditions for repairs and temperatures people had to deal with at night were starkly different. Widespread twenties and thirties made not having heat an uncomfortable situation. Shelters and hotels were packed with those seeking warmth. Neighborhoods were filled with the rumble and hum of generators (many of them newly purchased after Irene!), and lines for gas were brimming with folks trying to keep them running.

As is often sadly the case, the effort to stay warm turned deadly. A NOAA report linked thirty-nine deaths to the storm, both direct and indirect. Some of those were people who froze in their homes, and others who succumbed to carbon monoxide poisoning due to improper generator use. The others were hit by falling trees or electrocuted by live wires.

To help with the monumental task of cleanup, the National Guard was called in. Many roads were completely impassable, in some cases for days, with trees down at every turn. A nest of pine, oak, and maple choked back roads all over southern New England and parts of the tristate area.

Just getting down a street to start any sort of power restoration was extremely difficult and tedious.

Between the cleanup and the electrical issues, there was a plethora of early season snow days to be had nearly two months before the winter solstice even arrived. Many schools had to close for several days in a row thanks to lack of power and lack of safe conditions for buses to pick up children.

For those kids and their parents, the biggest trick of all was coming. Towns across the region cancelled Halloween. Or at the very least, moved it to a later date. Where it wasn't postponed, curfews asked that all trick-or-treating take place before darkness fell. It was just too dangerous.

Power lines were still strewn about. Windows were darkened by the lack of electricity. Branches and limbs were either laying in the streets or piled high atop the snowbanks. Many people simply just weren't home because they were in shelters or staying with family somewhere else. This was no way to run a night of fun for the kids. A whopping 86% of the northeast had snow cover on Halloween morning, which you sometimes do not even see during the heart of winter.

Dozens of cities and towns moved festivities for some time between November 2 and 7. This led to some tactical decisions being made on the part of savvy trick-or-treaters. If your town cancelled Halloween, you could go a town or two over where it was still on, grab your candy bars, and then double-dip with your own town the following week.

The attempt to move or cancel Halloween was met with quite a bit of blowback from parents who had kids anxious to head out. Who ever heard of moving a holiday? It would end up being a recurring theme in the coming years with Hurricane Sandy in 2012 and a couple of large late-October windstorms in 2017 and 2019.

In a final ironic twist, "Snowtober" was essentially the only winter weather that visited the region that year. It barely snowed again the entire season and was one of the warmest winters on record. It was as if Mother Nature played all its cards for one huge, freak event. It was the icing on top of a severe year of weather. From a tornado super-outbreak in April to Hurricane Irene in August. An earthquake even rattled the east coast before Irene came ashore. Massachusetts had its own large, destructive tornado in June. In all, there were fourteen separate billion-dollar disasters in the US during 2011.

Epic Snowfalls: The Great Snow of 1717 and the Snow Blitz of 2015

Announced by all the trumpets of the sky,
Arrives the snow, and, driving o'er the fields,
Seems nowhere to alight: the whited air
Hides hills and woods, the river, and the heaven,
And veils the farmhouse at the garden's end.
—RALPH WALDO EMERSON

THERE IS SOMETHING UNIQUE ABOUT WINTER'S ABILITY TO CHANGE A landscape. While some storms are simply known for tearing down trees or chewing up beaches, a fresh snow can completely transform the world around into something new. Barren pastures and stands of stark, leafless trees put on a shiny new white coat and life throughout takes on a new feel and rhythm in the wake of a storm. The small ones are picturesque. The medium ones are, perhaps, just right. Then there are the big ones. Singular storms that are disruptive and shut down the routine of our lives. Those do not come around every year but are frequent enough to be part of the conversation of New England winter. And on rare occasion, when the atmospheric combination lock unleashes a streak of successive blizzards, truly jaw-dropping scenes can emerge. Of all New England winters, two such barrages stand above the rest. The Great Snow of 1717, and the Snow Blitz of 2015.

Getting crushed with numerous snowstorms in a row is not common in New England. To illustrate that, these two events are nearly three hundred years apart and without equal. As you can imagine, the impacts between them were quite different considering society has changed dramatically over the course of a few centuries. We have thousands of videos, photos, charts, sharp satellite images, ground reports, and more to document 2015's epic run of snowfall. For the colonial period, it is a more anecdotal collection of observations and writings from diarists and authors of the time. Details for the latter are, naturally, a little fuzzier. Though general agreement among several accounts gives us a clear enough picture of a wild and memorable winter. Let's start with the old, then move on to the new.

In a normal winter, just December of 1716 on its own would have been something to remember. And it isn't even part of the Great Snow's main event! Winter rapidly rounded into mid-season form across New England with some five feet already on the ground in Massachusetts before New Year's Eve. To put that into perspective, during modern continuous record keeping since the late 1800s, the most December snow recorded in Worcester, Massachusetts, is 37.0 inches and in Boston 27.9 inches. Easy to see December of 1716 alone was way outside the norm for colonial settlers just trying to hang on and survive in a new world.

During January snow kept on coming, yet this too was not the heart of the Great Snow. Sidney Perley wrote in his 1891 book *Historic Storms of New England* that snow fell "in considerable quantities several times during the month of January, and on February 6 it lay in drifts in some places twenty-five feet deep." Yes, this is STILL before the worst of winter came to roost.

If that is the case, when exactly is the "Great Snow" period? Dates differ among accounts, so it is tough to say with precision. Everything we know is based off surviving written word, and weather is nothing if not subjective. Maybe a six-inch snow is a significant storm to one observer, but it has to be over a foot for another. We do know that there weren't many breaks, and it was gray and snowy throughout the month of February 1717. In Perley's rendition, a big storm is noted to have occurred on February 18 and that is the earliest cited starting point for the "Great Snow" period. "The great storm began on February 18 and continued piling its flake upon the already covered earth until the twenty-second; being repeated on the twenty-fourth so violently that all communication between houses and farms ceased."

Cotton Mather, an American Congregational minister in Boston who kept an extensive diary, made note of this particular storm but not until February 20. In his writings, the progression sounds more like a moderate snowfall that thickened the snowpack to start, followed by a much stronger nor'easter on the 24th.

Mather writes with the flourish of a minster indeed when describing this more potent storm. "But on ye Twenty-fourth day of ye Month . . . Another Snow came on which almost buried ye Memory of ye former, with a Storm so famous that Heaven laid an Interdict on ye Religious Assemblies throughout ye Country, on this Lord's Day, ye like whereunto had never been seen before."

A more current translation would say that there was one heck of a nor'easter burying the city and that church was cancelled. In deeply Puritan New England, the prolonged cancellation of religious gatherings was certainly a mark of how difficult times had become with an unmanageable amount of snow covering roads and homes. Local Native Americans who had achieved an age of nearly a hundred years old said they had never seen or heard of such a snowy period before, nor had their fathers.

Alas, weary New Englanders were not in the clear. Several more storms brought additional snow into the area through early March. From historical writings they are generally thought to have hit on March 1, 4, and 7. Altogether, the run of tremendous snowfall lasted from February 18 through March 7 and brought at least forty inches of snow to Boston and more than seventy inches for inland areas. The snow depth across northern Massachusetts was recorded at a depth of eight to sixteen feet deep after the final storm!

Many homes were completely buried by drifts up to twenty-five feet high, identifiable only by wisps of smoke curling out of chimneys. A disorienting landscape without the typical markers of roads and homes made finding friends and loved ones a monumental task. Any one-story home became an igloo with those trapped inside merely praying to be found. One account describes a widow trapped in her Medford, Massachusetts, home for several days. When searchers found a smoking snowbank, they dug in and revealed the home. Inside the widow had been keeping her children alive by burning the furniture, and her situation was likely not unique across the region.

Tunnels were made and maintained so that townspeople could find their way to other homes and farmers could reach their barns. A heavy workload in a day before plows and snowblowers.

Business and correspondence came grinding to a halt. Mail in 1717 was delivered along several post roads with navigation of these roads possible only with snowshoes, strenuous effort, and the passage of time. Snow drifts on the road from Boston to Portsmouth, New Hampshire, reached a height of fourteen feet. "Post boys" recruited to do the job blazed trails by way of snowshoe in late March. Even then, the forty-mile trip from Salem, Massachusetts, to Portsmouth took nine days on foot.

Much of colonial life was dependent on agriculture, and to that end the effect was devastating. Damage to orchards and subsequent fruit crops was extensive. The harsh conditions and heavy snow cover killed innumerable creatures both domestic and wild. Deer took the brunt of it, with an estimated 90% of the population being killed off over the course of the winter. Starving bears, wolves, and foxes became aggressive in their pursuit of livestock with nightly raids on sheep and chickens. Hundreds of cattle were buried and killed by the relentless winter, some of them found in the spring frozen and standing right where they were buried.

Cotton Mather also noted the devastating loss in his writings, noting that vast numbers of cows were destroyed. But there were also tales of the remarkable resilience of life in the face of adversity. He remarked on the incredible findings of live animals emerging from the snow weeks after the storms—"For no less than eight and twenty days after the Storm, the People pulling out of the Ruins of above an 100 sheep out of a Snow Bank, which lay sixteen feet high, drifted over them, there was two found alive, which had been there all this time, and kept themselves alive by eating the wool of their dead companions."

Poultry seemed best equipped to survive while trapped under the heavy snowfall. Live chickens were found after being buried for a week, and turkeys for as long as twenty days! One account even claimed that two pigs emerged a full twenty-seven days after the final storm, apparently surviving by eating a plant similar to an aster (a tansy).

And yes, there were even tales of love to emerge from the snow-bound situation. A cheeky quote from Joshua Coffin's history of Newbury, Massachusetts, reads "love laughs at locksmiths and will disregard a snowdrift." Coffin recounts a young man named Abraham Adams and his love, Abigail. Apparently, newlyweds as of December 1716, they had been separated by the storm with Adams living several miles away from Abigail and her family. A week had passed since the last storm and, well, natural instincts took over. Adams strapped on the snowshoes, made the three-mile trek over snow-covered hills, and made his way into her second-story chamber window.

News anchors often report (with a grin) a boom in "blizzard babies" after stormy times nowadays, and it is not a new concept. Baby Adams was born nine months later on November 25, 1717.

To set the stage for this extreme weather event, we again find a possible relationship to volcanic activity. While the "Year Without a Summer" was put in motion by the explosive eruption of Mount Tambora in Indonesia, the winter of 1716 to 1717 came on the heels of several powerful eruptions around the world. Most notably, Mount Kirishima in Japan, Kelud in Indonesia, and Taal volcano in the Philippines. The combined influx of ash and other particulates into the atmosphere likely blocked a portion of the sun's energy from reaching the earth, and the world was still in the period known as "The Little Ice Age" to compound matters. It would not be a reach to say the considerable volcanic activity helped to shape a colder and stormier winter than usual across North America.

Flash forward three hundred years. Life is a little different. We see snowstorms and harsh patterns coming well in advance. Animals can be protected, plows clear the streets, trucks cart snow away to "snow farms" instead of letting it hang around on the sidewalks. That said, there is less room to let the weather breathe. While the population of New England was somewhere in the area of 150,000 in 1717, it is now close to 700,000 in the city of Boston alone. And weather, well, is still weather. History loves to repeat itself and you do not need a series of volcanic eruptions to conjure up a highly unusual weather event. During the winter of 2015 a string of storms unlike any other the region had seen since those colonial times marched up the coastline. Over the course of four weeks from late January through late February, eastern New England was the new Siberia.

From a meteorological standpoint, the most significant difference between the two winters were their first halves. We do not have accurate records to know for sure how much snow fell November through April during 1716 to 1717, but I think it is safe to assume that it was more than what fell in 2014 to 2015. The amazing part of 2015's Snow Blitz was that it came out of nowhere, and it struck fast. The region had barely had a winter until the snow started falling in late January, with nothing more than a few inches before the parade of storms rolled into town. Christmas Day reached the sixties in many spots and some New Englanders spent it walking on the beach. It was looking like a dud of a season.

The snowfall total through the third week of January in Boston was 5.5 inches. People were getting cozy. Maybe it won't be so bad after all this year! As the fence lines disappeared and then the sidewalks and then the street signs over the following few weeks, no one with a long memory will ever feel comfortable after a slow start again.

The first salvo came on the evening of January 24. Not so bad, even festive. About a half foot floated down and many were just happy to finally break out the sleds and boots and enjoy winter. The perfect little mid-winter snow though just an appetizer for the multi-course feast that lay ahead.

On the 26th, a real storm was brewing. The government shut down non-essential services and instituted a travel ban. That night and through the 27th, the swirling snow intensified into a raging blizzard, plunking two to three feet of snow on the region. So much pounded on down that it currently stands as the largest snowstorm on record for Worcester, Massachusetts, with 34.5 inches and sixth largest on record in Boston with 24.6 inches Just like that, many towns and cities were close to their average for a full winter season. The good news was that, unlike in past storms decades or centuries earlier, word got out and everything shut down in advance of the storm. There were no deaths and very few drivers stranded on the roadways. At the time it was dubbed "The Blizzard of 2015" because, hey, how many blizzards do we get a year? Often not more than one, if any.

Alas, it did not take long to deliver more. On Groundhog Day, February 2, Punxsutawney Phil proclaimed six more weeks of winter and another snowstorm blanketed eastern New England with one to two feet of snow. At this point, life was starting to get tricky. The drifts were starting to push up to the base of windows on many homes and schools were cancelling class left and right. The back-to-back nor'easters broke records in Boston for the snowiest five-, seven-, and ten-day periods on record. A whopping 40.2 inches fell in a week, which crushed a previous record by nine inches. The pipeline of storms was not empty yet, however. Just a few days later, meteorologists were calling for another blizzard to strike.

The nor'easter of February 7 to 9 threw down another one to two feet of snow across eastern New England and added another Top 10 snowstorm to the list in Boston. It came in at seventh, with 23.8 inches of snow. Now, there was real trouble. Roofs were collapsing and structural damage became a hazard. The city of Boston, famous for quaint cobblestone paths and tight spaces, was crippled with snowbanks blocking movement and leaving mounds on the sides of streets faintly resembling the shapes of the cars buried underneath.

People were laughing to keep from crying. Generations of New Englanders had come and gone and never seen something like this. There are winters where several big storms strike, though usually separated out by a little time or at least with some melting between the events. Not during the blitz. Constant cold made for powdery snowstorms and a total lack of melting to make room for the next one. Boston went fifteen days without ever reaching the freezing mark during the blitz and forty-three straight days without reaching 40°F (a new record). February would wind up as the first or second coldest month on record for numerous northeastern towns and cities. It made for a preposterous, once-in-a-lifetime sight. Tunnels to get to homes, gigantic icicles hanging everywhere as snow dams mounted, and first-floor windows succumbing to the snowbanks. Neighborhoods looked reminiscent of those on the Tug Hill Plateau in New York, well known for feet upon feet of lake-effect snow burying everything in sight. And then came the coup de grace. Yes, another blizzard was on the way. The proverbial cherry on top of a bombogenesis sundae.

And after all this, what better time than Valentine's Day? Anyone with a deep passion for winter and snowstorms had a lot to love on February 14 as nature decided to add a final exclamation point and thrust the winter of 2015 into the annals of weather history.

The final storm made plenty of noise and became famous for the intense amount of thundersnow that lit up the coastline on the morning of February 15. Once again, one to two feet of snow fell across eastern New England. While more flakes would fall before the year was done, this was the end of the blitz and it made for unprecedented amounts of snowfall for such a short period of time.

The real epicenter of the historic run was Boston, and the records set during this window may be out of reach for a long time to come. In just thirty days ending February 22, 94.4 inches of snow fell. For context, that is more than every single *entire season* recorded in the city since 1891, except for two others (1993 to 1994 and 1995 to 1996). It obliterated the most ever witnessed in a month before with nearly two feet more than second place. It was even thicker across Downeast Maine, where 132.5 inches fell in Eastport over five weeks' time. The city of Bangor also notched its snowiest month on record.

The sheer amount of it in such a short period of time was mind-boggling and captured the attention of both national and global media. After a few more inches made their way down in the weeks ahead, the record for snowiest winter ever recorded in Boston was achieved with a final tally of 110.6 inches.

A collection of iconic images will be remembered by locals for decades to come. Huge chunks of ice floated across Massachusetts Bay and washed up along the arm of Cape Cod, most notably in Wellfleet Harbor. "Iceberg" tourists flocked down to see them in person. Half-frozen waves, which affectionately became known as "slurpee waves," lapped at the shoreline during the extended cold. In Boston, workers diligently trucked snow out of city streets and sidewalks every night for weeks and created a massive snow farm on an empty South Boston property. The result was something that looked like it came out of a movie—a snow pile so big that king of the mountain was not played by schoolchildren, but by excavators and backhoes. So immense was this pile (the apex reached seventy-five feet high) that it stood the test of time for months. The final shred of filthy leftover ice and snow melted on July 14.

We give part of the responsibility for 1717's Great Snow to volcanic activity. So what happened in 2015? In the modern case, an extremely anomalous pattern emerged across western North America and stayed stuck in place for nearly two months. A huge ridge extended across western sections of North America, while a deep trough in the polar jet stream took up residence across the eastern half. This unusually strong dipole kept a constant flow of arctic air going from Siberia straight to the eastern US. The remarkable part of it all was the persistence. The cold set records for duration, but in Boston not a single daily record low was set the entire time. The polar flow was also full of disturbances ready to explode on the edge of that arctic air, which banged up against warm waters along the east coast. The result was this exceptional run of powerful nor'easters that affected eastern New England more than any other part of the country, and a winter none will forget.

Catastrophic Ice Storms

"Limbs just kept on falling and falling and falling. At night, you heard the snapping, crashing, banging of limbs. You expect a storm to go on at night and then end. But the length of this storm was overwhelming."
—BILL BERGEVIN, LANDSCAPE ARCHITECT, BATES COLLEGE

IF YOU ARE LOOKING FOR SHEER DESTRUCTIVE POTENTIAL IN THE WINter, look no further than an ice storm. Covering the land in a heavy, frozen blanket they can be both mesmerizing and deadly. No other weather event is quite like it, and if you've experienced a bad one there is no forgetting the sights and sounds. The gunshot-style bangs during the night as branches snap. The eerie bluish glow of the sky due to arcing transformers. And the next morning, an awe-inspiring scene of sunlight glinting off forests and towns encased in ice. These types of storms, more than any other, look like something out of another world. When they get out of control, the damage can be otherworldly as well.

There are two relatively recent events that have plunged millions into the dark and produced extreme damage across the northeast. In 1998, an epic and seemingly endless ice storm over several days featured such a heavy coating that it crumpled high-voltage transmission towers and took away electricity for over four million customers. Canada, New York, and northern New England took the brunt. Then in 2008, another extreme event focused on interior southern New England with a smaller but high-impact footprint. Here we look at what makes a setup ripe for these costly and destructive storms.

What exactly is freezing rain, anyway? It is one of the most confusing concepts when it comes to winter weather. If it is 26°F at my house, then why isn't it snowing? To answer this, we must look up and think about the whole column of air above our heads. Picture yourself in a hot-air balloon that can travel tens of thousands of feet high. On a typical weather day, you can hop in that balloon and expect the air to get colder the higher in altitude you travel. And indeed, hiking a mountain will give this same impression! You are comfy at the base but need extra layers at the top.

However, that is not always the case. Sometimes there are inversions, which mean the air above is warmer. Inversions are frequent on calm nights where the surface cools off rapidly. Another way to sneak warmer air above is by advection. Essentially, storms pushing air masses around. Warmer air is less dense than cold air and will happily take the path of least resistance. It rides up over the top of cold air instead of pushing it out of the way. This is how an ice storm gets its start.

Now that the warm air is flowing in, it takes control of a layer thousands of feet up where snowflakes would typically be falling during a winter storm. Instead, you get a feed of air that is just too warm for snow to develop in or survive. Water droplets dominate and head toward the ground as rain.

Here comes the trickiest part—how cold is it below? Cold air is stubborn and very difficult to kick out, especially in valleys where it is easiest for warmer air masses to zip right over the top of. These bowls in the terrain can hold on to their supply of subfreezing air for a long time. If the cold layer is too deep, then the rain will have a chance to freeze before it hits, which we then call sleet. A little icy pellet pings off everything and is tricky to drive on, but other than that is fairly benign when it comes to damage. To get a major freezing rainstorm, there must be a shallow but consistent layer of subfreezing air. Just enough to bring the temperature of the rain to or even a bit below freezing (called supercooled water droplets) so that it is ready to immediately freeze on contact.

In January of 1998, the setup could not have been more perfect. Heading into the winter, one of the strongest El Nino events on record had come to life. A warming of the equatorial Pacific Ocean, El Nino winters are known for strong subtropical jet streams that bring large amounts of tropical moisture eastward into the US. The Great Ice Storm of 1998 was a high-precipitation event, and the strong subtropical feed was a critical ingredient to fuel this long-duration ice storm. It takes two to tango, and to produce a dangerous amount of frozen precipitation there would have to be cold air to meet that tropical feed. That too was on the way.

The surface analysis on January 5 would reveal a sprawling 1036 millibar arctic high across the southern tier of Canada. A stationary boundary separating this cold air mass and the milder air to the south formed, setting up near the border between Canada and the Great Lakes. Temperatures were largely in the twenties north of the front, and as light rain developed it froze immediately on surface objects. At this first stage, the icing was primarily in Canada and Maine.

By the 6th, high pressure was consolidating in Quebec and the flow of cold air began to intensify. Instead of staying locked up across Canada, it now began to ooze south through the Saint Lawrence River Valley of Quebec into the Champlain Valley of Vermont and New York. To the east, the state of Maine increasingly gave up ground to the cold being forced in from the northeast. As the shallow arctic air gained a better foothold, overrunning precipitation continued. Rain was increasingly turning to ice on trees and power lines in the subfreezing air.

The next day, all the ingredients had come together to put a region into the throes of an historic event. What was going on higher up in the atmosphere was in direct opposition to the surface, flooding the mid-levels with mild air and moisture but keeping cold air in control near ground level. A large ridge in the jet stream had built over the eastern US, allowing for warm and moist Pacific air to rise and produce heavy precipitation. The arctic high anchored over eastern Canada kept a fresh flow of cold surface air in the twenties draining into New England and New York. Nothing seemed to want to budge, and so every bit of rain that fell continued to accrete on all surfaces it met. From the evening of January 7 through the next two days, it continued without reprieve. A series of smaller waves of low pressure rippling along the surface boundary made for constant freezing rain, which is what made this ice event stand out among all others. One singular storm would have come and gone with lesser and more manageable ice totals. In 1998, some towns had freezing rain for over a hundred hours straight.

Under normal circumstances, ice building up to a radius of a half inch is a significant ice storm. Branches start snapping, the power begins to go out, and driving is impossible. Just that half an inch can add five hundred pounds of extra weight to a power line. These more minor events are still a big deal and can be quite disruptive. During the 1998 ice storm, a half inch is what built up during the first couple of "primer" days in the Saint Lawrence and Champlain Valleys.

The deep layer of tropical moisture, wrung out in escalating fashion January 7 to 9, brought more precipitation than what is usually seen in an entire month, and almost all of it froze. Across northern Vermont and the Adirondack Region of New York, four to six inches (liquid-equivalent) were observed for the event. More than half of it grabbed everything in an icy embrace, with storm total ice accretion reaching two to four inches. Massena, New York, recorded a maximum of 3.5 inches with 115 total hours of freezing rain and drizzle. Since reliable records for freezing rain duration began in 1948, the duration was nearly double the second-longest event. It was an eye-popping amount that became an unstoppable force.

Catastrophic damage developed swiftly. The resulting scene was a war zone of blocked roads and devastated forests. Trees came crashing down and power lines sagged toward the ground before snapping. The electrical grid took an historic blow unlike anything experienced before in the region. Entire transmission towers for electricity succumbed to the great weight of increasing ice. In total, 130 of these were destroyed and more than thirty thousand fell. Over four million customers quickly lost power, the majority in Ontario and Quebec but a significant 1.25 million across the northeastern US. In the difficult mid-winter conditions that followed, it would take weeks to get lights back on in the hardest hit areas.

In Maine, nearly 80% of the population lost power and schoolchildren who were on Christmas break had their holiday break extended by up to two more weeks. Making matters more complex were the variable conditions found across the state. A major ice storm for many in central and southern Maine, over a half foot of sleet in the middle of the state, and up to two feet of snow across northern Maine! Governor Angus King (later to become state senator) declared a state of emergency and deployed the National Guard to clean up wreckage and supply water. Thousands of storm refugees made their way to emergency shelters, as weeks without power in the heart of winter can easily turn deadly. Six Mainers would die as a result of the storm.

The immediate risk to life was not necessarily the ice clinging to everything, but the need for heat. Being northern New England in the dead of winter, it was frigid and not having heat in the home is dangerous. Many locals in this part of the world are well equipped with woodpiles and pellet stoves to get through the tough times. Generators are a useful commodity as well but can be a risky tool. Many of the estimated forty-four deaths (US and Canada) were indirectly related to the storm in this way. Carbon monoxide poisoning from improperly used or installed generators, fires from space heaters, and injuries attempting to clean up the large number of damaged trees accounted for many of lives lost. Several who were unable to heat their homes perished from hypothermia.

In the forests, it was mayhem. Across New York, Vermont, New Hampshire, and Maine there were trees in absolute disarray falling into each other and across roadways, making them impassable. So far as total acreage goes, Maine fared the worst. An area ten times the size of Rhode Island, eleven million acres, was damaged or destroyed by the Great Ice Storm of 1998. Behind Maine came New York with 4.6 million acres of damage and New Hampshire with just over 1 million acres. Beech trees and sugar maples fared the worst, which had a lasting impact on the maple syrup industry for years to come. Farmers saw widespread damage in their orchards as heavy ice ripped apart thousands of fruit trees.

A Forest Service survey after the storm detailed the layers in which the ice storm was most prolific and reveals the challenges of forecasting such events. There are usually only thin zones where the atmospheric profile is just right for significant icing. The survey found that in New York, the most damaging ice buildup occurred at an elevation below 1,800 feet. But in Vermont, there were two slices. One being in the Champlain Valley below 1,000 feet and a secondary slice in the Green Mountains between 1,800 and 3,200 feet. The city of Burlington recorded an extensive ice accumulation of 2.2 inches. To the east in New Hampshire, it was the White Mountain region south to Mount Sunapee across elevations of 1,000 to 2,600 feet. In localized zones people were dealing with catastrophe while just outside those bands the impact was much less significant.

Downed trees, lack of power, and impassable roads proved extremely costly for dairy farmers who had no way to milk their cows or get their product out. Milk was dumped out in vast quantities. For New York State alone over $4 million worth of milk was wasted. For other farmers the loss was of a more immediate nature when barns collapsed under the weight of ice and killed livestock inside.

Without question the ice takes top billing for this storm due to the sheer magnitude of area affected. Flooding, however, was also a part of the problem. Since there was a tremendous amount of rainfall from January 5 through 10, water that wasn't freezing up immediately was released straight into rivers and streams. Over four inches of rain, combined with ice jams, caused widespread flooding across New York and Vermont. Evacuations were needed to get over a thousand people to safety. The Black River Valley in New York was among the hardest hit.

After millions of destroyed trees and a critical loss of infrastructure, the Great Ice Storm of 1998 came with a price tag of $4.4 billion between the US and Canada. Bad as it was on the US side of the border, it is considered one of the worst natural disasters to ever strike Canada. Montreal saw a crushing 3.9 inches of ice accumulation with freezing rain and drizzle reported for eighty hours. For the area most affected, it was handily the worst ice event since at least the ice storm of 1929.

TEN YEARS LATER . . .

A lesser but still extremely damaging ice storm came back to New England ten years later, but this time it was the southern half that would experience the dramatic sight of a world on ice. From the Capital Region of New York eastward through Massachusetts, southern Vermont, and New Hampshire, there was a more localized but impressive event.

Unlike 1998, this storm struck with more speed instead of a drawn-out affair. On December 10 many would not have expected a winter event of significance was about to start. High temperatures reached into the fifties and sixties across southern New England. It was not built to last as a late-day cold front ushered in fresh delivery of arctic air. It was a tenuous air mass since the timing was early in the winter season, but cold enough at the lower levels for mixed precipitation. More importantly it was very dry, which would allow temperatures to stay below freezing by way of evaporative cooling once the rain arrived. That afternoon the National Weather Service issued an Ice Storm Warning calling for the potential of "widespread and extended power outages."

During the daylight hours of Thursday, December 11, mostly plain rain fell with just light icing across Massachusetts, New Hampshire, and Vermont. To the casual observer, there were no glaring signs of impending overnight destruction showing just yet. An area of low pressure developed across the southeast and moved to the coastline, tracking over the Mid-Atlantic on Thursday and then over southeastern Massachusetts by Friday morning. It was to the west of the center that cold air remained locked in place near the surface with northeast winds draining subfreezing air from northern New England.

Even though it was moving a lot faster than the 1998 edition, there was still a significant plume of tropical moisture streaming up from the Bahamas. Intense rainfall picked up Thursday night and lightning lit up the sky in some of the heavier downpours. A drenching of more than two to four inches of rain came down in a short amount of time. This made the 2008 icing unique. Normally it is difficult to get a significant ice storm when it is pouring outside. Ice builds up more efficiently when the event is spread out over time, or if the surface air is unusually cold (more toward the mid-twenties). Heading into the overnight hours the coldest surface temperatures being reported were just a degree or two below freezing. Nevertheless, as the sun set and solar insolation no longer played a role, a quick but hard-hitting ice storm picked up speed.

When night fell, so did the trees. Without the sun to hold icing at bay, freezing rain started to build up in earnest. Across the higher elevations of central Massachusetts the calls started coming in shortly before 7 p.m. Rutland, Holden, Leicester, and the area around Wachusett Mountain were among those experiencing an extreme deterioration of conditions. Transformers arced and lit up the night, mingling with the occasional flashes of lightning. The air was full of the sounds of cracking trees ringing out like shotgun blasts and their collisions onto cars and homes. The town of Paxton's Public Works Director, Mike Putnam, called the destructive chorus of the night "one of those things that doesn't go away" in your memory. It was so dangerous to attempt clearing the roads that he was forced to call all the crews back until the next day to get a full grasp on the situation.

Ice storms have the interesting quality of being beautiful . . . until they're not. At dawn, the scene was shocking. Anyone who had gone to bed with minimal ice during the evening woke up to a completely different world of blocked roads and wrecked trees in all directions. It was not the epic two to four inches of ice accretion that had crushed northern New England ten years previous, but a half to one inch of ice had gripped power lines and branches. That was more than enough for considerable carnage.

Overnight there were over a million customers from Pennsylvania to Maine who had lost power. The majority were in Massachusetts (350,000) and New Hampshire (440,000). For hard-hit New Hampshire, it was the largest outage in state history. Public Service of New Hampshire (PSNH) called the storm unprecedented in terms of damage and had to rebuild entire sections of the grid. Closer to the cold air source, Maine was once again experiencing widespread significant icing. Between the 1998 and 2008 events, Maine would end up dealing with the most overlapping damage. Power outages were the greatest since the 1998 storm and numbered 220,000 customers.

The need for electricity and heat was the main headline and would be a source of vocal disapproval in the days to come. People living in storm shelters or hotels for days on end became extremely frustrated and utility companies came under fire for the slow response. A main point of contention was related to the issue of location. Much of the severe damage occurred in tight clustered areas and anyone living outside the icy zones had no idea such a catastrophic event had played out. In most backyards across southern New England, it was just wet. You almost had to go looking for it to know how bad it was. The Monadnock Region of New Hampshire to northern Worcester County presented one extreme pocket of icing, with another in the Berkshires above 1,000 feet and another around the southern Greens of Vermont. In New York around Albany, it was the worst ice storm in twenty-one years. Though if you were just a few miles away from these prime areas, you would not see any ice at all. A drive up a few hundred feet and it was chaos.

It would take over two weeks for the final customers to receive power again. The provider Unitil, which provided service for northern Worcester County into New Hampshire, came under more scrutiny than most. Fines were recommended by the Attorney General of Massachusetts, though were never levied. However, Governor Deval Patrick signed into law the "Unitil Bill," which required investor-owned electric and gas companies to annually supply the state with emergency response plans. The company also dramatically increased tree-trimming programs, which would soon come in handy during major tropical-weather-related outage events in 2011 and 2012. Demanding more communication and preparedness from utility providers would become the lasting legacy of this damaging ice storm.

The 100-Hour Storm

"The crew shoveled, had coffee, shoveled, had lunch, shoveled, had a beer, shoveled and shoveled and shoveled and the beer supply was getting dangerously low."
—BRUCE SLOAT, A.M.C. HUTS MANAGER

MANY STORMS FOLLOW A SIMILAR SCRIPT. THERE IS A FORECAST, A buildup of anticipation, impact, and then they depart for us to pick up whatever mess is left behind. Every once in a while, a memorable storm doesn't really check all the boxes. These unique ones can have equally unique outcomes that go outside our usual expectations. Let's take a deeper look at one of these meteorological oddballs, which may be the biggest storm you've never heard of.

The "100-Hour Storm" in February of 1969 was no beast of bombogenesis or fearsome flood-maker. It did not shut down the megalopolis corridor. But it did produce snow. A LOT of snow. A plodding and relentless system, the swirling area of low pressure hung around the coast of New England like a crazy relative you wish would just finally leave on Thanksgiving.

It ended up setting records that have yet (as of this writing) to be broken and buried towns in an unreasonable amount of snow. Combined with another storm earlier in the month, it made February of 1969 an extremely memorable month for winter weather. In parts of New Hampshire, over one hundred inches would fall before it came to a close.

The seeds of the storm came from nearly nothing. A glance at the weather map on February 22 just showed a couple of weak troughs of low pressure—one in the Great Lakes region and another near the southeast

Atlantic coast. No digging polar jet stream to be found or arctic disturbances plummeting down from Canada.

On the 23rd, a weak area of low pressure would start to organize and
drift up the eastern seaboard toward the New Jersey coastline. Surface
pressure began to drop modestly, and by the morning of the 24th it had
fallen to 994 millibars. This is a pretty run-of-the-mill area of low pressure by northeast standards. Certainly nothing noteworthy in the land
of rapid pressure falls and dynamic coastal storms. For the rest of its life
cycle, it would barely budge from this mark with the lowest observed
pressure coming in at 992 millibars. Home barometers? Unimpressed.

Benign as it was by this metric, the snow started falling with enthu-
siasm across southern New England on the 24th. Likely unbeknownst to
those on the ground, a four-day storm was underway.

Despite the lack of classic ingredients, there were a couple things
working in favor of big-league winter weather at the time. Players on the
field that certainly would pique a forecaster's interest. For one, it was an
El Nino winter. El Nino is a phase of a natural process in the equatorial
Pacific Ocean, called the El Nino Southern Oscillation (ENSO). In an El
Nino year, this region off the South American coast and near the equator
is much warmer than usual. El Nino and February in New England are
like peas in a pod. In many of these types of winters, the season can get
off to a slow start during December but produce some big storms later on.
February of 1969 was no exception.

The other favorable pattern at the time was a strong negative phase of
the North Atlantic Oscillation, or NAO. A negative phase of the NAO
means that there is a large area of blocking high pressure in the vicinity of
Greenland. It is not necessary for a major northeast storm, but it can help
out by slowing down the weather pattern and holding cold air in place.
The US and Canada were full of slow-moving, meandering systems at
the time. These "blocky" patterns often give the east coast a great shot at
blockbuster winter weather. Sure enough, this is precisely what happened
with the 100-Hour Storm. There was nowhere for it to run.

From February 24 through 28 it would painstakingly move past Nantucket and then swirl east of Cape Cod. While it was merrily spinning
away, a sprawling area of high pressure anchored around Hudson Bay in

Canada was draining cold air southward and balancing out the milder marine influence of the Atlantic. It perfectly set up for some long-lasting snow bands, the heaviest of which would form over Maine, New Hampshire, and Massachusetts.

What better place to look for huge piles of snow than around one of the snowiest spots in the eastern US? Mount Washington goes by the slogan "Home of the World's Worst Weather" and quite often lives up to the reputation. Standing 6,288 feet high and peering down upon the Presidential Range of the White Mountains, it can see snow even in the middle of summer. In February? It is game on. Even still, this storm would deliver more than anyone there had ever observed before.

"The winter of 1969 was a winter, that if you survived it, you did not forget it" recalled Bruce Sloat, Huts Manager for the Appalachian Mountain Club at the time. Stationed in Pinkham Notch at the base of Mount Washington, this was tall praise for a place with high winter standards. During a 2003 retrospective in Mount Washington Observatory's newsletter *Windswept*, Sloat took a look back at this peculiar event.

Even below the lofty peaks, he noted the winter started with a bang. November brought a record snowfall with twelve inches in a single day and thirty-five inches for the month. December proceeded to stack up 66.5 inches and at this point everyone was just in good spirits for a successful ski season ahead. It snowed on two-thirds of the days in January, bringing another 31.2 inches. Topping the century mark before the snowiest time of the year was impressive to say the least. The snow depth was seventy inches as February began, and the stake used for measuring required a four-foot addition to keep track. In most years, that would do the trick. Not this time. The biggest of all was coming.

At Pinkham Notch the 100-Hour Storm started slowly, with just some flurries on February 24. Then came day two with twenty-one inches. And then the next day with 24.5 inches. And the next? Another twenty-seven inches. Sloat notes that at this point, even the plows were crying uncle. It was nonstop work just trying to keep up. The snow depth had increased to 164 inches by the evening of February 27. Or if you prefer a visual, standing about as tall as the largest African elephant you can find.

For the staff living there, this meant snow piled well above the first floor of all structures.

At last, after a four-day barrage, the skies relented. A more reasonable 4.5 inches fell on the final day, February 28. The total stood at seventy-six inches of snow; the greatest snowstorm the area had ever recorded. Tunnels were just about the only way you could find the visitor's center or any other building in the vicinity. Avalanches large and small were common, including one that slid off a roof and broke through a window of the Pinkham Notch Trading Post. But no one complained about the skiing conditions! Markers on the trees had to be placed fifteen to seventeen feet up and all the underbrush was completely covered in snow anyway.

This was all occurring at an elevation of 2,031 feet. What was happening higher up as all that moist Atlantic air rose up the slopes of the Presidentials? Hearty souls of the Mount Washington Observatory were witnessing history themselves. The constant stream of snow bands traveling west from the ocean brought a biblical amount of snow that has not been replicated since.

In October, perhaps one of the observers had an inkling of what the future held. Whitney Barry entered into the log on Halloween, 1968, "A windy, cold day. We received 19 inches of snowfall during October. If all this keeps up it should be quite a winter!"

The Chief Observer, Guy Gosselin, shared the same line of thought. Gosselin said he had a pretty good idea of what kind of winter it was going to be by the end of December, when the observatory had already notched more snow than twenty-one of the thirty-six *full seasons* on record atop the summit.

Up here, previous high marks were smashed by the 100-Hour Storm. A tongue-in-cheek entry in the Observatory's log on February 24 starts with "A long overdue snowstorm is nosing up the coast and promises to be interesting."

Over the course of four days, 97.8 inches of snow would blanket the summit. Perhaps most impressive was the new record for greatest twenty-four-hour snowfall; 49.3 inches falling on February 25 alone. With no trees and intense winds constantly scouring, most snow typically blows down into the gulches and valleys leaving less impressive amounts on the rockpile. In this case, drifts over ten feet high remained, which were tall enough to walk on up to the roofs of several summit buildings. The seasonal breakdown on Mount Washington is seen below with the 100-Hour Storm leading the way in an obscene stretch. February 1969 remains its snowiest month on record and it was the snowiest season overall. No other before or since has even reached five hundred inches.

Monthly snowfall on Mount Washington, 1968 to 1969

September 1.5 inches
October 19.0 inches
November 86.6 inches
December 103.7 inches
January 56.8 inches
February 172.8 inches
March 95.0 inches
April 21.1 inches
May 5.6 inches

For observers on the peak of a mountain, it was meteorological magic. On the ground where people lived though, the extreme snowfall was dangerous. In Berlin, New Hampshire, the Notre Dame Arena's roof collapsed from the weight of all that snow. The timing was terrible with a hockey scrimmage between Notre Dame High School and Berlin High School about to begin. Players were on the ice taking practice shots from the blue line when the roof came down, and tragically a fifteen-year-old boy was killed. Norman Boucher, a goalie for Notre Dame, was the sole casualty of the collapse with several others injured. Hundreds rushed in with shovels to help when a call for help was put out over radio.

Numerous other roofs collapsed in the region, including a warehouse operated by Brown Company. The town of Berlin had to declare a state of emergency twice that February, and the *Berlin Reporter* described how townspeople had to tie ribbons to the antennas of their cars so that others might have a chance to see them over the huge snowbanks at intersections.

While New Hampshire took the brunt of the storm, there was a wide footprint of heavy snow. Long Falls Dam, Maine (near Sugarloaf

Mountain), reported 56 inches, and Old Town, Maine (just north of Bangor), came in with 43.6 inches.

The cities were not spared, either. Much of eastern New England was socked with days of snow. Portland, Maine, had 26.9 inches just two weeks after being crushed by the February 8 to 10 blizzard with 21.5 inches. February of 1969 is the city's second snowiest month on record, just behind January of 1979.

Impressive totals continued down the coastline into Massachusetts. An astonishing thirty-nine inches was reported in Rockport, Massachusetts. With Rockport sitting on the tip of Cape Ann, it usually isn't able to stay cold enough for such a prodigious amount of snow. Truly a memorable event for an oceanside town.

The city of Boston observed over two feet—26.3 inches. At the time that was good for the largest snowstorm on record before the Blizzard of '78 topped it and then later on the Presidents' Day Blizzard of 2003. Official observations noted snow for 101 consecutive hours at Logan Airport.

Fortunately for New York City, the worst of it stayed off to the east. Mayor John Lindsay was battling a firestorm of criticism from his handling of the blizzard that occurred from February 8 to 10 (that is, oddly enough, much more famous than the 100-Hour Storm). This preceding nor'easter is often referred to as "The Lindsay Storm" and will come up in an online search much more readily. A forecast of rain and sleet turned into fifteen to twenty inches of snow, and the mayor's handling of the event nearly cost him his career. If the next storm had hit, at least they would have been prepared! Lindsay sent out hundreds of workers at the mere sight or sound of snow for the rest of the season.

Without truly being a blizzard and having no major coastal impacts via wind and waves, how does this event compare to some of the "classics"? Well much like hurricanes and tornadoes, there is a scale that ranks snowstorms called the Regional Snowfall Index, or RSI. Guess who owns the top spot over more than five hundred other storms in the database? The 100-Hour Storm.

Considering how many historic, epic blizzards there have been over the decades, you may be surprised to hear this. So how could this lesser-known one rise to the top?

The scale works like this. All events are analyzed by the National Centers for Environmental Information (NCEI) to answer three main questions. 1) How big of an area did it cover? 2) How much did it snow? 3) How many people did it affect? RSI, most importantly, is also based on the climatology of a particular area. Or in other words, it takes into account what's typical for the region you are looking at (a major snowfall in the southeast is less common than one in New England, etc.). The scale features categories 1 to 5, with 5 being the most severe.

The 100-Hour Storm from February 22 through 28 comes in with an RSI value of 34.026, a Category 5 storm that is just ahead of the March 1993 Superstorm for top honors. The biggest reason for this is due to the highly unusual snow totals over a large population. For instance, 7.7 million had at least twenty inches fall in their town and over two million people lived in an area that saw over thirty inches. It is very rare to cover that many people with that much snow! In the end, that is what makes this storm an infamous one.

As winter came to a close, fears were running high that the story would end with a destructive round of spring flooding. Nature decided to give everyone a break instead after the wild winter. The 100-Hour Storm was the last major snowstorm of the season with just minor events to follow, and then warmer weather came gradually. The quieter and milder progression into spring produced an orderly melt that released water from the snowpack without too many issues. Mountains turned green again and everyone affected had plenty of stories to tell and look back on before the next winter came knocking.

PART TWO
Summer in New England

July 10, 1911, saw a record high of 105 degrees in North Bridgton Maine.

IF WINTER IS A MARATHON, SUMMER IS A GIANT PIZZA AND BEER WAIT-
ing at the finish line. The reward for making it through to the other side.
Unlike the blistering Great Plains or sweltering Gulf Coast, New England summers are generally pretty tame and enjoyable. Fire up the grill,
admire the blooming hydrangeas, and soak it in. If anything, vacationers
end up grumbling about coastal fog or rainy stretches more than uncomfortable heat. However, that is not always the case.

When the jet stream packs its bags and heads north for the summer
season, the spring rains often fade to drier and hotter times. Peak sunshine in both length of daylight and intensity can bake lawns and farms.
If the atmospheric faucet turns off and thunderstorms become scarce,
then before long drought and fire can enter the dialogue. Drought lends
itself to hotter temperatures, and suddenly the backyard BBQ plans turn
to a search for relief.

Make no mistake, a New England drought is not a California, or
Midwest-type affair. Massive wildfires burning hundreds of thousands of
acres of land are not a staple of this region, nor long stretches of 100°F
heat. Many inland towns across southern New England average around a
dozen 90°F days per year. At the coast or with elevation it is in the single
digits. Air-conditioning was not even widespread in New England until
the past couple of decades because it wasn't worth the trouble. A bedroom
fan could get you through most of the time.

So when out-of-the-ordinary heat sets in, the effects are significant.
All weather is relative, and no one is equipped to deal with a long duration
of extreme heat in this part of the world. Someone hanging out at the pool
in Texas won't have a visceral reaction to a stretch of nineties for weeks on
end, but a New Englander will break out in sweat once the mercury pushes
80°F. Hitting 100°F calls for a regional shutdown until the heat breaks.

Even in an age when summers continue to grow hotter overall, the
greatest heat wave to strike New England came over one hundred years
ago. Go back another one hundred years and you will find a year where
summer failed to take shape and sweaters were favored over shorts. Before
modern improvements to firefighting and monitoring arrived, summer
flames ignited across the region and burned towns to the ground. While
summer often conjures images of cookouts and seaside vacations, a bucolic
season is not a guarantee.

The 1911 Heat Wave

"By the fourth day, a pattern has emerged: awake to the heat, work in the heat, seek relief from the heat, toss and turn in the heat."
                                                    —The Hartford Courant

The deadliest natural disaster to strike New England? It was not a monstrous hurricane, raging flood, or mile-wide tornado. It was heat. Simple, sweltering, insufferable heat. For eleven days in 1911, a summertime inferno of historic proportions engulfed the region.

The front page of the *Bangor Daily News* on July 3rd summed up the developing situation:

Bangor People Just Broiled Alive.

It wasn't too much of an overstatement. At the end, there were estimates of two thousand people killed in New England. It could be more, but we'll never have an exact count. We know at least 1,100 certificates showed heat as cause of death in Massachusetts alone. Outside of New England, many hundreds perished to the south and west across Pennsylvania, New Jersey, and New York.

Some committed suicide or, in other cases, went insane. As many as twenty ended up under watch in an asylum. To avoid such a state of delirium, there was a natural rush toward the closest body of water around. But even that search for relief was not safe. At least two hundred people died by drowning in local ponds, lakes, and the ocean.

Tracking the toll was, and still is, difficult. It is not as simple as trees falling or cars crashing. There is a myriad of ways heat can adversely impact the body. Heat stroke, heart failure, dehydration, exacerbating preexisting conditions, etc. Do drowning deaths count if they are related to an attempt to escape the heat? To this day there are disagreements as to whether heat or cold kills more people due to the difficulty of attribution. Regardless, the loss was extreme and never again repeated in the region.

For days on end a sweaty sea of humanity lay strewn about towns and cities. On benches, under any tree or sliver of shade they could find, or rooftops at night. Some children were even spotted playing in birdbaths. Those who could get to the beach made a beeline. A *Boston Globe* report estimates that two hundred thousand showed up at Revere Beach, a long sandy haven lining the cool Atlantic waters north of Boston.

In a more strenuous age, getting work done became nearly impossible. Many factories roasted like ovens and had to send workers home. While "Neither snow nor rain nor heat nor gloom of night" allegedly slows the postal service, this brand of heat was too much. Mail delivery was briefly halted for much of the region. Roads turned gooey as tar melted and bubbled. In Worcester, Massachusetts Police Chief David Matthews hired a motorized vehicle to help bring the stricken to local hospitals. The typical horse-drawn ambulances could not keep up with the number of people needing assistance.

We all expect the hot midday sun to be severe, but particularly in the tight quarters of cities the nights were dangerous as well. To answer the call, several local mayors opened the parks for sleeping, and the masses took them up on the offer. Thousands across the region slept under the stars, turning places like Boston Common into one huge uncomfortable slumber party with an estimated five thousand guests.

Even at dawn there wasn't much improvement with temperatures staying above 70°F for five straight nights. After a brief break, yet another five-night stretch above 70°F moved in. This has become something much more common in recent times, but it was completely out of character in 1911.

Wherever people could find a breath of fresh air, they gathered like a moth to flame. In one sad story shared in the *Hartford Courant*, a twenty-eight-year-old man named John Merlo was sleeping on the tin roof of a boardinghouse in the tenement district of Hartford, Connecticut. There were few other options for some relatively comfortable rest.

In the middle of the night, boarders on the ground level were awoken by a loud noise. When they looked outside, they found Merlo dead on the ground with a snapped neck. He had rolled off the roof in his sleep.

All of these tragic stories took place over an eleven-day stretch; the historic temperatures coming in two acts with a quick intermission. During an age before any sort of accurate advance forecasting, no one could have seen it coming. June was pleasant, if not a bit cool. There wasn't anything out of the ordinary in the weather department.

On a national scale, the seeds had been sown. The National Oceanic and Atmospheric Administration, or NOAA, has a useful way of tracking droughts called the Palmer Drought Severity Index (PDSI). This index takes into account precipitation, temperature, and soil moisture data. It is best used to show water deficits over fairly long periods of time, roughly six to twelve months.

The PDSI in July of 1911 showed that everyone east of the Rockies was in rough shape. Why does it matter? When the land is dried out, more of the sun's energy is spent heating up the ground and air with less energy spent on evaporation. This can become a feedback process, and in turn you will find many of the nation's hottest all-time temperatures are found during times of dry weather/drought.

Which brings us back home to New England on July 2. Temperatures spiked and climbed their way into the low and middle nineties. On July 3, an escalation. Boston, Massachusetts; Hartford, Connecticut; and Concord, New Hampshire, all sizzled to the triple digits. Even Burlington, Vermont, made it all the way up to the century mark. This, on its own, was a rare occurrence.

And then the 4th of July came calling. Naturally, a day of celebration and a typically rowdy day to bask in the glory of our nation's birth. Instead, nature brought its own fireworks to the significantly subdued party.

Declining to gather around a hot BBQ, many people made the move to crowd around local thermometers just to see just how high they would go. Nashua, New Hampshire, rocketed up to 106°F, the hottest reliably measured temperature during the onslaught. Boston simmered to 104°F, the hottest temperature on record there and it is still standing over a hundred years later.

Your typical "heat wave" in the northeast is generally defined as at least three straight days reaching 90°F or higher. Many cities and towns left this benchmark in the dust. Concord, New Hampshire, reached triple digits for three consecutive days. In Boston, the century mark was notched three out of four days.

After the holiday, patience and fortitude were being tested. Hot weather in a place where extended high temperatures are relatively infrequent can be a slow, silent, and inescapable menace. Looking at extreme weather events of the past, you do not see many gripping photos of the 1911 heat wave. Storms produce damage, snowfall iconic images of drifts and ocean waves, but heat's main mark is left on the human body.

Everyone was feeling that drain on July 5 after several days of sizzling sun, but no immediate help was offered from the heavens. The day came and left just the same and baked under more record temperatures.

It was the following morning when the *Meriden Morning Record* gave a heat wave halftime report. Readers picking it up were on to day five of oppressive weather and were greeted by a rather apocalyptic front page.

FIVE HUNDRED ARE DEAD,
ALL RECORDS BROKEN BY MEMORABLE HEAT WAVE

GIRLS DROWN IN ICE POND

MUCH DAMAGE BY TERRIFIC THUNDERSTORM

These were trying times. But that very day, after topping out near 100°F again—relief! Sweet relief. Though not the quiet kind. The atmosphere finally boiled over and sent a destructive line of severe thunderstorms across New England, knocking down trees and causing damage throughout the region. Several more were killed by the storms, adding to the heat wave's toll.

In Bangor, Maine, this was a particular insult to injury. An enormous fire had burned the city two months previous. The heat and then the storms came as an unwelcome encore. Windows were smashed, bricks torn off buildings, and an outdoor entertainment tent hosting a Wild

West–themed show ripped apart. The *Bangor Daily News* called it the worst storm damage there since 1883.

The only saving grace was that behind the thunder came an opportunity for bodies to cool and minds to focus. Highs in the seventies and eighties allowed for a brief respite. Unfortunately, it wasn't over yet.

By the 9th, the heat wave was back on. This final salvo brought another four days of nineties, with some cities hitting 100°F one more time (varying by location with occurrences on both the 10th and 11th). To put this into perspective, the city of Concord, New Hampshire, has only hit 100°F nineteen times between 1869 and 2019. Four of them came in 1911, and no other year has had more than two. It was a true outlier of extreme weather.

After one more round of severe storms, the run was over. The 13th brought an end to the inferno and major heat almost completely avoided the region for the rest of the summer. Boston only reached 90°F three more times for the remainder of the year, all of them in August (a summer month that also featured eight days that never even made it to 70°F). But the damage had been done.

It is hard to imagine such an impact in modern times, and it would be accurate to say we'll never see anything like it again. The heat? Absolutely possible that a stretch of such temperatures could be equaled in the future. Perhaps even the near future. But the impact on lives cannot be duplicated thanks to technology. You are more likely to spend a day complaining about being too cold in an office building during a hot July day now than risk injury or death from the heat.

Back then? There was no escape. No air-conditioning. No electric refrigerators. Or for that matter, electricity at all in some cases. No trays of ice cubes to throw in your water or tea unless you could grab an elusive chunk from an ice wagon rolling through town. No large cooling centers open to the public, no backyard swimming pools. And of course, no reliable long-range forecasting to let everyone know it was coming.

Flash forward to now and the warning would be given over a week in advance that a period of intense heat was coming. The grid would prepare, outdoor events would get cancelled, strenuous activity avoided, precautions taken with extra water and misters at public gathering spots. We are fortunate to have such advances to avoid the misery and loss of 1911.

It also has to be mentioned that the style of the day was less than helpful. Looking back at photographs of the heat wave, you may be struck by the clothes. In suffocating heat, many are seen wearing hats, jackets, suits, and full-dress attire. New Englanders were a modest bunch and apparently thought covering up properly was worth risking death. People would be generally half-naked walking around town in such heat nowadays.

There is one other big factor to consider—the changing landscape. New England countryside and cities alike were much less forested back in the early 1900s than they are now. Between the land clearing for farmland in the 1800s and city expansion, the lack of trees is glaring in photos from the time. The cities were full of dust, metal, brick, and stone. All those surfaces soaking up the heat of the day and slowly releasing it at night (what meteorologists refer to as the Urban Heat Island).

Anyone who has spent time outside knows the refuge a leafy canopy can provide or how much cooler the forest feels versus an exposed city street. There were fewer of these places to escape the glare of the hot sun.

The urban epicenters for heat could be considered even worse in 1911 than now in a way, considering the inadequate ventilation and tight living quarters that many (particularly the poor) had to suffer in. The statistics clearly show how amplified the mortality was in cities versus the outlying areas. Unofficial temperature readings at street level, in the brick and stone jungles of cities, were over 110°F.

All of this is of course of little solace to the people who tried their best to cope with the 1911 heat wave with few resources at hand to help. Even children of the time had a long time to wait. Air-conditioning in residential homes did not arrive until the 1930s (for the wealthy) and did not become common until the 1940s.

The Year Without a Summer, 1816

"To be alive in the years 1816–18, almost anywhere in the world, meant to be hungry."

—GILLEN D'ARCY WOOD, PROFESSOR,
UNIVERSITY OF ILLINOIS AT URBANA-CHAMPAIGN

SWEATING OUT JULY AND AUGUST WOULD COME AS NO SURPRISE TO anyone, but how about shivering your way through summer? Trading T-shirts for overcoats and swim trunks for mittens. Imagine watching lakes freeze and snowflakes float by your window. Fruits and vegetables withering in fields or on the vine. It sounds like some sort of nuclear winter or sci-fi thriller. But in 1816, summer was truly a season that never existed in New England.

It became known as "The Year Without a Summer" or the year "Eighteen Hundred and Froze To Death" to many in the northeastern US and Europe. Crops died, prices soared, people starved, migrations to new lands got underway. 1816 and the two cold years that followed may have been the most significant meteorological event to affect the world in the nineteenth century.

What sort of phenomenon could conspire to erase the warmest time of the year? At the time, no one on the ground living through it knew. Was in punishment from an angry God? The sun burning out? News in the early 1800s did not travel any faster than the fastest boat at hand. Today, we know the main driving factor was bubbling in the tropics on the other side of the world. A volcano, with all its explosive power, ready to blow. When it did, the results were immediate, catastrophic, and felt all over the globe.

Mount Tambora is a large stratovolcano on Sumbawa Island, Indonesia. Like many volcanoes, it has had numerous eruptions in its lifetime. None of them are anything like what happened during April of 1815. It is widely considered the largest volcanic eruption in recorded human history, packing ten times more power than the violent and deadly Krakatoa eruption nearly seventy years later.

It started with tremors and pyroclastic flows on April 5 as the beast awakened. Tambora sputtered lava and shook the earth for several days until the main event. In present day the volcano stands 9,354 feet high, which is impressive in its own rite. In 1815, it stood nearly a mile taller. The eruption was so intense that more than four thousand feet of the mountain were blown to bits during its cataclysmic eruption on April 10, 1815.

Ten thousand people were immediately killed. Pyroclastic flows rocketed across the landscape at speeds of over a hundred miles per hour, tsunamis rocked the Java Sea, and falling debris wiped out everything in the vicinity. Lava boiled the ocean and pumice rock choked harbors for miles around. Ash fell for weeks, standing up several feet thick in spots and burying towns. It rained down as far as eight hundred miles away and accumulated one centimeter deep 550 miles away in Java. Tambora threw 150 times more ash into the sky than Mount Saint Helens did in Washington State during its famous 1980 eruption. In the months ahead, eighty thousand more people would die across Sumbawa and surrounding islands due to disease and famine because crops simply could not grow.

Volcanoes have a scale for intensity called the Volcanic Explosivity Index (VEI). Like earthquake magnitude, it is logarithmic and ranges from 0 (non-explosive) to 8, which is deemed a "mega-colossal eruption," capable of ejecting 240 cubic miles of solid material. Tambora was rated VEI 7, nearly as strong as it gets on earth. The sheer force sent plumes of smoke and gas twenty-eight miles up into the atmosphere, which is where its effects around the world start to take shape.

To cause a notable shift in earth's climate, both a volcano's ability to launch particles into the stratosphere and its location matter. Any volcano can throw ash and smoke high into the air, which blocks incoming solar radiation and can produce a short-term, localized cooling. But these larger particles usually fall out of the sky over the course of a few days and the effect dissipates. Only the most explosive eruptions are able to send smoke and gas more than ten miles up into the stratosphere. If you are going to create a global event, this is where the dirty work happens.

The most important substance to make it this high is sulfur, and Tambora spewed plenty of it. An estimated sixty megatons made it into the atmosphere. Once sulfur reaches the stratosphere, it starts to mix with water vapor and the result is sulfuric acid aerosols. These small droplets, in turn, produce a hazy film in the air that blocks some of the sun's rays from reaching the surface. Unlike the heavier pieces of ash that quickly fall out of the troposphere below, this haze can linger for months or even years. A shroud falls over the surface below.

Powerful volcanoes are found all over the world, but they're not created equal in their potential for a period of global cooling. If a major eruption takes place in the higher latitudes, like Iceland or Alaska, the aerosols ejected into our sky tend to stay in the higher latitudes. But volcanoes in the tropics have a much better potential for light-blocking power to spread. As stratospheric winds carry them all over the planet, the dimming can have a much greater reach. If conditions are right, the aerosols head into both hemispheres. The worst impacts in the aftermath of Tambora remained across the northern hemisphere, where land temperatures dropped roughly 3ºC in its wake.

This brings us back stateside to New England, where unbeknownst to everyone a natural disaster was about to deliver much confusion and suffering. Though the Tambora eruption occurred in April of 1815, the results on the ground hadn't been felt yet in the US. The winter of 1815 to 1816 wasn't notably harsher than others. Then spring came, and winter never left. Every single month of the year ended up featuring a frost.

In May, fruit trees started to bloom. Then a series of heavy ice and frosts killed buds and emerging flowers. There were days when temperatures tried to rise and restart the growing season, but every time it was met with a crash and frustration. Crops would start to grow and then a heavy frost would return. Trees couldn't get their internal engines going and remained barren, looking more like November or December. In our entire tree ring record, 1816 is the only year where a growth ring can't be found.

June, the month of flowers, began with a promise. A very warm day in the nineties was met with optimism on June 5. Which was promptly followed by plummeting temperatures and a snowstorm. The 6th and 7th delivered a nor'easter that dropped a half foot of snow on parts of New York and New England. After the storm, the snow depth in Cabot, Vermont, was recorded as a staggering eighteen inches deep. It is believed, though perhaps less reliably so, that flurries were even spotted in Boston on June 7. That is the latest account of snowflakes seen in the city any time during its history since settlement.

New Hampshire Governor William Plumer had the inauguration ceremony for his second term on one of these cold, bleak June days. He used it to tell people that the weather situation was "God's judgment on Earth" and that people should humble themselves for their transgressions.

Similar sentiments were shared by Rhode Island's Governor, William Jones. He issued a proclamation designating a day of public "Prayer, Praise, and Thanksgiving" while making note of the alarming amount of illness.

In Connecticut, a clockmaker named Chauncey Jerome recorded his thoughts, which were later published in an autobiography. He called the summer of 1816 a time "which none of the old people will ever forget, and which many of the young people have heard a great deal about."

Jerome noted that there was snow and ice on at least one occasion during every month of the year. In contrast to the 1911 heat wave that scorched 4th of July festivities, the 1816 celebration was spent bundled up. Jerome wrote "On the 4th of July I saw several men wearing thick overcoats while the sun shined on them. A body could not feel very patriotic in such weather." The 5th of July came with reports of frost.

There aren't concrete daily weather records for this time like we have now, but observant townspeople and scientists did keep diaries and journals. There were reports of lows near 30°F in northern New England and 40°F as far south as New Haven, Connecticut. Lake and river ice was seen as far south as Pennsylvania during this time.

It is unfathomable to picture ice on the 5th of July in southern New England, but it speaks to the power of geological forces and how small we are in the face of them. They can render our norms useless in a heartbeat.

Looking back, perhaps it is easy to understand anyone who thought it may be the end of times. Given the state of the landscape, it must have seemed like it.

Williams College professor Chester Dewey observed that "the trees on the side of the hills, whose young leaves were killed by the frost, presented for miles the appearance of having been burned or scorched. The same appearance was visible through the country—in parts, at least, of Connecticut—and also, on many parts of Long Island."

The Reporter in Brattleboro, Vermont, gave their account of the situation in its July 17, 1816 issue.

"The Season—It is believed that the memory of no man living can furnish a parallel to this present season. From every part of the US, north of the Potomac, as well as from Canada, we have accounts of the remarkable coldness of the weather, and of vegetation retarded or destroyed by untimely frosts. In Montreal, on the 6th, 8th, and 9th of June were falls of snow, and from the 6th to the 10th, it froze every night. Birds, which were never before found except in remote forests, were then to be met with in every part of the city, and among the [flocks], and many of them benumbed with cold, dropped dead in the streets."

A particularly odd part about the weather was its irregularity. The cold wasn't firmly in control. There were still days that soared into the nineties and even a few reports of triple-digit heat. Soon as it would turn a corner, a rapid drop would move in. On top of the temperature ups and downs, it was extremely dry. The drought made growing conditions even worse, and even sparked numerous fires late in the season that burned fields and darkened the skies.

The erratic nature had farmers in complete fits. The end of August featured a killing frost. In fact, the frost reports dipped as far south as Virginia on August 20 and 21. Then the extremes continued right into September with a late-month snowfall in parts of New England. Apples and corn were destroyed in large quantities. Over a half of the entire corn crop was rendered useless. The price of oats skyrocketed from twelve cents a bushel to ninety-two cents. Some farms had to kill off their livestock because they did not have enough to feed the animals. Plus, they needed food for themselves!

At this time in human history, much of life depended on the harvest of any given year. A series of bad years could lead to starvation or migration, and indeed huge numbers of people made their move after the failed growing season. Tens of thousands hitched up a wagon and made the trek west. Ten percent of Maine's residents departed for warmer climes.

The westward expansion of the US can thank, in large part, the Year Without a Summer. Farmers in New England already had a tough enough job in a fickle climate. 1816 was too much to bear, and so they decided to try their fortunes somewhere else. Midwest towns and states quickly swelled in the years ahead.

Author Samuel Goodrich commented on the failing will of the time, writing that "In the pressure of adversity, many persons lost their judgement, and thousands feared or felt that New England was destined, henceforth, to become part of the frigid zone."

In an odd twist of history, we may be able to thank the terrible weather for Mormonism. In Norwich, Vermont, the family of Joseph Smith Jr. was battling failed growing seasons. His father finally threw in the towel in 1816. They made the trip west to Palmyra, New York, where it was believed farming conditions would be easier. Smith was ten years old at the time and would go on to found the Church of Jesus Christ of Latter-Day Saints in New York fourteen years later.

While struggles were significant across the northeastern US, the same period also featured widespread famine and disease in Europe. Cold temperatures, heavy rainfall, and flooding destroyed crops and led to a major typhus epidemic. In Ireland alone, it is estimated that one hundred thousand were killed. There were food riots in Britain and France while Switzerland declared a national emergency due to the lack of food and civil unrest. Germans called the following year, 1817, "The Year of the Beggar." This was all going on in the aftermath of the Napoleonic wars, which had just ended in 1815.

One positive piece of pop culture emerged from Europe during these dismal times. An eighteen-year-old Mary Shelley and her future husband happened to be visiting Lord Byron at his villa in Switzerland. The weather around Lake Geneva was frankly too awful to enjoy the wonderful countryside, so a lot of time was spent indoors. Lord Byron would come up with ways to pass the time, like writing and sharing stories. He challenged all in attendance to write a ghost story, which is precisely what Mary Shelley ended up doing.

The night before her ghost story was "due" to Lord Byron, she had a dream of a man who was not a man, but a monster. One made by a scientist playing god. The inspiration that caught her that night in the middle of drab, cold, and stormy summer gave birth to the story "Frankenstein."

In the end, all the hardship seen in the years from 1816 to 1818 cannot be completely blamed on the eruption of Mount Tambora. Earth was still in a period referred to as the "Little Ice Age" and had already been experiencing a period of cooling with harsh winters and poor growing seasons. This colder stretch of climate lasted roughly from 1300 to 1870, during which time there was a general increase in volcanic activity (in addition to Tambora) and also several stretches of low solar activity.

It is understood that a lack of sunspot activity correlates with a slight decrease in solar output, and there were several minimums during the Little Ice Age. The Maunder Minimum was the deepest, lasting from 1650 to 1710. During the time of "The Year Without a Summer" the sun was in a period we refer to as the Dalton Minimum, named after English meteorologist and chemist John Dalton. It is possible that low sunspot activity may have affected global circulations at the time, and in conjunction with volcanic cooling could have been another factor leading to bouts of famine and unseasonably cold weather in parts of the world.

It was a teachable time as we try to unlock all the factors that warm and cool the globe. Climatologists can easily spot the Tambora eruption within ice cores extracted from Greenland and Antarctica by way of their very high sulfur content. Using it, another signpost pointing the way toward the history of earth's climate puzzle. The more pieces we have, the better we can predict extreme weather in the future.

The Year Maine Burned, 1947

*"Only chimneys and foundations of houses, and twisted iron of stoves
and plumbing and tools remain—silhouetted like weird distortions
against a pall of smoke that covers the land and reaches a fog bank far
out to sea."*

—THE BOSTON GLOBE, OCTOBER 22, 1947

BLINDING SNOW AND RAGING TORRENTS OF WATER CAPTURE THE IMAGI-
nation and leave their mark on us all. Sometimes, a lack of storms can have
the very same effect. The absence of any interesting weather can breed
danger, as was the case in the autumn of 1947. With endless warmth and
little rain, New England dried up and flames soon followed. It would culminate in the worst natural disaster in Maine's history; an inferno scorch-
ing 206,000 acres of forest and wiping out nine towns in a matter of days.
All thanks to stormy weather taking a month's long hiatus.

"The Year Maine Burned" certainly did not start out that way. There
is a running joke that the seasons in Maine are winter and construction.
The cold season is very long and most everyone looks forward to breaking
free by the spring. That, too, can be a tall task in Maine. March offered a
break as mild and bright weather depleted the winter snowpack. But after
a temporary reprieve the spring of 1947 was gloomy and wet, providing
little relief from the winter chill. At the time, it was the wettest April
through June on record in Caribou with 13.64 inches of rain (and melted
snow) coming down. That record would continue to hold on for the next
sixty-four years until falling in 2011. Talk of drought and wildfire was far
from anyone's mind. Mainers just wanted some sunshine.

Nothing is built to last in the world of weather, and the prevailing pattern made a 180-degree turn toward the end of June. All of the

sudden it was summer and the regional thermostat bumped up considerably. Multiple days managed to soar into the nineties across the Pine Tree State and precipitation dwindled. For outdoor enjoyment, it was the opportunity anticipated by all during the first half of the year. Reason to celebrate! However, every action has consequences and at the same time the forests were drying out.

Autumn arrived and still the summertime weather endured without much change of pace. Between the start of July and October, the Palmer Drought Severity Index (PDSI) had gone from "moderately moist" to "severe drought," a rapid transformation of conditions on the ground. Some locations across the state had gone 108 straight days without measurable rainfall, and you could see it on the trees. Leaves began to change color and fall to the ground prematurely, crispy enough to crumble by hand. Now the natives were getting an inkling that, if luck turned on them, trouble could be down the road.

There is no lack of wildfire fuel across Maine. Approximately 90% of the state was covered in forest in 1947, a rich mix of pine and deciduous forest. The coniferous trees were experiencing a "cone year," which meant they were producing an abundance of seed-filled cones efficient at spreading flames. On top of this mildly unsettling development, the forest floor was still littered in slash, or leftover debris, from the Hurricane of '38, which had ripped down large swaths of trees. As the calendar turned a page, the soaking gales of October failed to materialize. Signs of danger were flashing brighter and the landscape became a tinderbox.

It wasn't just the available fuel or the missing rainfall adding to this mix, but a run of impressive late-season heat. New Englanders had been soaking up one of the most memorable "Indian Summer" stretches in memory. You'd be hard-pressed to even know it was autumn. Over the border in Concord, New Hampshire, there had barely been a day under 70°F during the month of October. There would end up being nine days where temperatures made it into the eighties, reaching a peak of 89°F on October 17. To the east in southern Maine, Portland notched seven such days. There hasn't been an October in the city before or since with more. In terms of average monthly temperature, the only October warmer than 1947 came seventy years later in 2017.

Now all it would take was a careless spark and the situation was not going unnoticed. With risk escalating the Governor of Maine issued a state of emergency and closed woods to campfires, hunting, and general travel. Fire watch towers that generally close for the season by the end of September were re-opened by the Forest Service to keep a lookout. The public was warned to keep open flames to a minimum and to have their chimneys cleaned.

How does a cataclysmic fire begin? Before settlement most wildfires were started naturally by lightning. With increasing population came the rise of human-caused fires by way of careless flame. Humans now cause the vast majority of wildfires, whether accidentally or intentionally by way of arson. It was a building concern that any small spark could erupt and grow into a destructive blaze.

Indeed, by early October, reports of several small fires were popping up across Wells, Portland, and Bowdoin in southern Maine. By October 16, at least twenty separate fires were burning. But it was the next day, October 17, that would kick off the climax of Indian Summer with five straight days in the eighties away from the immediate coastline. It would culminate in the worst disaster the state had ever experienced.

In one of the most picturesque and peaceful locations in the US, a serious and destructive fire was about to ignite. Mount Desert Island (MDI) is the largest island in the state of Maine and second largest on the entire eastern seaboard behind New York. Home to Acadia National Park and the bucolic town of Bar Harbor, it would seem like an unlikely spot for Maine's worst natural disaster to begin. Nevertheless, the phone rang at the local fire department at 4 p.m. of October 17. It was a Mrs. Gilbert, who lived near Dolliver's dump on Crooked Road. She was calling to report smoke rising from a nearby cranberry bog. This was not unusual or unexpected. It had been dry and small fires were likely to occur, though the cause of this one was unknown. For three days, it did not present itself as a major threat as it burned up a mere 169 acres. Then on the 21st, the wind picked up.

Fires were increasing during mid-October, but it was the arrival of strong winds that turned a manageable situation into a statewide catastrophe. On MDI, they increased to over fifty miles per hour with peak gusts as high as seventy miles per hour. Immediately 169 acres became two thousand acres and a call for help was made to aid the outmatched local firefighters. The Army Air Corps, Navy, Coast Guard, University of Maine forestry program, and Bangor Theological Seminary all joined in to try to contain the blaze.

Swirling winds would prove too much to overcome as firefighting efforts were constantly shifting focus to save nearby communities. It continued to spread and on the 23rd it was Hulls Cove that seemed to be in the crosshairs, albeit briefly. That same afternoon, a cold front swept through the region and the wind changed its mind yet again. Now, it was Bar Harbor directly in line for the still growing wildfire.

Gale-force gusts quickly directed the fury toward town, traveling a distance of six miles in just three hours and burning a path three miles wide in the process. As word got out, a complicated and frantic evacuation got underway. Bar Harbor residents fled by whatever means necessary. Many who had run to the town's wharf scrambled to any vessel available and the resulting scene was referred to as the "Dunkirk of Bar Harbor." Hundreds joined the flotilla to escape by sea, and it must have been quite a sight to look back from the ocean to see such a gorgeous place engulfed in smoke and flames.

For an overland alternative there is only one road that goes in and out of MDI, and fire blocked it. Hastily activated bulldozers answered the call and were able to clear debris on Route 3 by the evening. A mass exodus by terrifying car ride through sparks, ash, and flame was the exit strategy for two thousand residents. The evacuation was largely successful given the short notice. Only five were killed, two suffering heart attacks and two dying in a car accident. The other was a man who went back for his cat and was never seen again.

For several generations dating back to the mid-1800s, Bar Harbor had become a summertime playground for the rich and famous. The swanky retreat set along the spectacular coast of Frenchman Bay had annual visitors such as the Rockefellers who owned palatial estates. Their stretch of Bar Harbor was known as "Millionaire's Row" and featured some of the finest homes in New England. The inferno was indifferent to their pricey property and torched sixty-seven of these seasonal estates. Damage to the mansions alone was over $8 million in 1947 dollars and many of them were never rebuilt, changing the face of the town. Instead of sprawling retreats, they were replaced by motels in the years after the fire to encourage tourism. In addition to the opulent summer cottages there were 170 permanent homes and five historic hotels razed to the ground. Unlike the high-priced luxury accommodations, most of the local homes were rebuilt. Miraculously the heart of the firestorm missed the central business district of Bar Harbor and spared many of the shops and commerce.

As it continued a path to the sea, the fire scorched Jackson Laboratory before running out of room to burn. In all, 17,188 acres were burned on Mount Desert Island, more than ten thousand of which were inside Acadia National Park. Rain would not arrive until October 29, and the flames were not completely out until heavier rain and snow fell in November. The official declaration that it was over was made at 4 p.m. on November 14 nearly a month after it had started.

So great was the destruction that you can still see the burn scars today by climbing one of the high points such as Cadillac Mountain. There was at least one positive as a result in Acadia though, and that was a change in the composition of the forest. What was mostly coniferous spruce and fir trees in 1947 now includes a blend of hardwoods that took root after the destruction. You can trace areas where the fire raged on a crisp autumn day as visitors take in the breathtaking fall foliage that would have been nearly all green before the 1947 fire.

The MDI inferno made global headlines, but it was not even the biggest fire burning in Maine at the time. When the winds reached their peak on October 23, numerous conflagrations exploded across the state. Four became larger than MDI's disaster. One to the north in the Centerville-Jonesboro area of Downeast Maine, which burned nearly twenty thousand acres, and three to the south.

Everyone had been on high alert watching for any hint of trouble. Even Bates College students were put on shifts as temporary fire wardens and firefighters. The women kept watch and the men ran out to join the fight on the ground wherever it was needed. One student, William Dill, wrote to his parents about the call to arms on the Lewiston campus.

"Last night almost all of us volunteered for fire-fighting duties. From suppertime on the fellows were vigorously discussing whether or not to answer the appeals which were periodically being made by the radio stations. Many left during the evening to report to firehouses for assignment, and at about 11 o'clock almost all the rest of us decided to go."

Dill, along with a group of forty other Bates students, were dispatched to the town of Richmond along the Kennebec River. On October 23, which would become known as "Black Thursday" in parts of the state, they were put to work as one fire spread toward the town center. He described the incredible disaster unfolding in front of the young men.

"It was spectacular to watch at night: a tree suddenly bursting into flame, a cloud of red and black smoke rising from a desolate blackness, and a stump burning like a candle. But a helpless feeling, too, was inspired by watching the fire bow and scrape to the whims of the wind. The ease with which things caught fire was almost unbelievable."

Farther south, the wind-whipped flames overtook entire towns across York County, which was destined to become the epicenter of a terrible week. There were several smaller fires that acted as starting points. The winds brought them together in what would become known as the Shapleigh and Kennebunkport fires, producing the most ravaged region in the state. The two account for over 130,000 acres of burned land. Of course, it wasn't just land. Devastation was swift and total in the communities of Waterboro, Shapleigh, Newfield, Lyman, Dayton, Alfred, Kennebunk, Biddeford, Arundel, Wells, and Saco.

This southern York County blaze became so immense that it had a perimeter 150 miles around, surrounding fifteen different towns. An unbroken fire front eight miles across roared toward the Atlantic, hopping over Route 1 and through Kennebunkport. Flames over one hundred feet high quickly razed forest and property straight to the coast. Many families had little more than a few minutes to evacuate as the Goose Rocks and Cape Porpoise sections of town immediately lost two hundred homes. What was left behind was made of stone. Monoliths of chimneys sticking up out of the ground amidst the smoke and haze. Leftover foundations provided markers where neighborhoods once stood. The central village did manage to survive thanks to a last-ditch effort. Firefighters bulldozed a fire break on the edge of the commercial district of Kennebunkport and were successful and protecting most of it. To the north Biddeford lost all but six homes at Fortunes Rocks.

One of the most fierce and destructive fires leveled portions of Waterboro and Newfield. Like at Mount Desert Island, the first reports of fire came in on October 17 near Russ Corner. It spread into Shapleigh and Newfield, and by the 20th Ossipee Mountain was in flames. Then the wind came on the 23rd, which steered the fire clear of the north side of town but blew it instead south and east across Route 202, torching the south side of Waterboro and wiping out fifty-eight homes. Everyone from the Red Cross to homeowners jumped into cars and attempted to speed out of the fire's path, all packing as many of their earthly belongings as possible and making a run for it.

For anyone who has not been in the heart of a raging wildfire and running for your life, it is hard to fathom the apocalyptic scene. The supervisor of forest fire control at the time, Austin H. Wilkins, tried his best in summarizing the pure hell that had engulfed York County.

"It was (an) awesome sight to see the solid walls of roaring fire sweeping over the mountain and across level areas, consuming everything in their path. The smoke hung so thick and heavy for days over the area that it was difficult at times to determine just how near or how far away the fires were. The sun did not penetrate through the thick wall of smoke for over ten days. Men using trucks, cars, and bulldozers had to use their headlights as much in the daytime as at night." The sound of the rushing fires, sometimes on fronts several miles long, was described as a continuous frightening roar.

The noise of it all stuck in the memories of those who lived through it, similar to a freight train as the winds urged the flames onward. A Waterboro man by the name of John Smith who had just come back from service in World War II told the forest service that it was as bad as anything he experienced in battle.

"I'd been overseas, and I think I was as scared during the fire as any time when I was over there, at times. You just figured that you weren't going to get out of it. Because you figured there was nothing that was ever going to put this fire out, you kind of were getting the feeling the whole state was going to burn. In fact, there wasn't much that stopped it until it got to the ocean."

The same frightening experience was playing out just to the northwest when another massive fire erupted in Brownfield, a town just south of Fryeburg and east of the New Hampshire border. Eventually chewing through 21,120 acres, it would become the state's third largest fire of the event and burned right up to the Oxford-York County line. Brave as they were, firefighting crews were woefully outgunned. Brownfield did not even have a permanent fire department. Just a frantically clanging church bell and a call for volunteers to rise to the occasion.

Much like in nearby Newfield and Waterboro, there was a procession of evacuees streaming out of town carrying pets, clothes, and whatever else was important to them. For hundreds, it would be all they had left by the end of October 23.

The inferno poured itself down over Peary Mountain and into the town, taking all but one home in East Brownfield and destroying the town's center. All of the buildings important to daily life were turned to ash, including the schools, post office, commercial district, and town hall. In neighboring Newfield, the entire town was overcome and burned.

Volunteers were rushing in from across Maine, New Hampshire, and Massachusetts to join the lines and try to save homes. An estimated twenty thousand firefighters, ranging from professionals and military to everyday citizens, threw themselves into the fray. There was heroism to be sure, especially considering the limited tools they had available and the rough terrain in which to work. Even cloud-seeding jets were dispatched to the area. But exceptionally dry fuels and strong winds were too much of a force to hold back. It was the return of wet weather almost a week later on October 29 or a fire simply running into the ocean that ended the outbreak.

When the month drew to a close, the Red Cross estimated that 3,500 people were displaced and 2,500 homeless. Across 35 communities, 851 homes were destroyed with an additional 397 summer cottages lost. Throughout the state sixteen people lost their lives, and that number certainly could have been a lot worse. Quick evacuations and the surge of volunteers to dig fire lines and open escape routes saved many from becoming casualties.

As is the case repeatedly with these extreme disasters, the Year Maine Burned was a catalyst for organizational and societal change to deal with the risk of wildfire. In the aftermath numerous towns across Maine established local fire departments. Volunteers were properly trained and standards were created for equipment and firefighting technique. Funding started coming in from the state and federal government, allowing communities to purchase trucks and modernize. The state of Maine also received a license for a two-way public safety radio network, something that was sorely missed during the 1947 firefight.

Just two years later, the Northeast Forest Fire Protection Commission was founded. Starting first with New England and New York and later including the Canadian provinces of Quebec and New Brunswick, it is a group founded on the idea of regional coordination and mutual aid. It still exists today with the addition of Nova Scotia, Newfound and Labrador, and Prince Edward Island.

The Dark Day of 1780

'Twas on a May-day of the far old year
Seventeen hundred eighty, that there fell
Over the bloom and sweet life of the Spring
Over the fresh earth and the heaven of noon,
A horror of great darkness, like the night
In day of which the Norland sagas tell,
The Twilight of the Gods. The low-hung sky
Was black with ominous clouds, save where its rim
Was fringed with a dull glow, like that which climbs
The crater's sides from the red hell below.
—John Greenleaf Whitter, Tent On The Beach

It is Friday morning during a time of renewal and spring weather, mid-May. The sun rises and fresh leafy canopies stretch toward the light. There is hope for a productive day ahead. But as the hours pass, something feels amiss. Daylight shifts gears into reverse. Instead of getting brighter, the sky appears to dim. In a nearby vernal pool, frogs begin to croak as evening birds join the chorus. Barnyard chickens and cows decide to pack it up for the day and head to their roosts and barns. By lunchtime, you can barely read a book without the aid of a candle.

On May 19, 1780, New England fell under a shroud of darkness and confusion. The "Dark Day of 1780" was not a product of an eclipse or great battle of the American Revolution. Nor was it punishment for the sins of townspeople from New Jersey to southern Maine, though I suppose we can't totally rule that out. A true cause was surmised but not scientifically proven for more than two hundred years. For those who lived it, a truly bewildering scene unfolded.

An unidentified poet would succinctly capture the events of the day.

> Nineteenth of May, a gloomy day,
> When darkness veil'd the sky/
> The sun's decline may be a sign,
> Some great event is nigh/
> Let us remark, how black and dark
> Was the ensuing night/
> And for a time the moon decline,
> And did not give her light

Sunrise that morning was noteworthy only for its brilliance. And perhaps also for its smell. An attention-grabbing shade of red surrounded the sun as it breached a curtain of haze on the horizon. Historian Sidney Perley of Salem, Massachusetts, wrote that the sun had risen a "deep, brassy red" and that a "strange and enchanting hue" robed the landscape all around. Even a man busy attempting to win independence for America, General George Washington, made note of it in his diary. While camped with the Continental Army in New Jersey, he described "heavy & uncommon kind of clouds—dark & at the same time a bright and reddish kind of light."

As for the scent in the air, there had been reports of a smoky odor for several days previous. There was conjecture that it may be from some distant fire, and there was more evidence a large blaze was occurring at some undisclosed location. On the morning of the 19th townspeople along the banks of the Merrimack River in Newburyport, Massachusetts, experienced a sooty rainfall and ash covering the ground to a depth of up to four inches. Much of it was full of burnt leaves, indicating a fire with either manmade or natural origins was somewhere nearby. Black scum was floating down local waterways.

It would not take long for the hazy sun to fade into inky darkness. In Vermont, it was already dimming minutes after sunrise. Then the shroud spread south and east, reaching Massachusetts by mid-morning. A Harvard Professor, Samuel Williams, placed the deepening darkness in Boston at approximately 10 a.m., writing: "This extraordinary darkness came on between the hours of 10 and 11 a.m. and continued to the middle of the next night." A day without daylight was underway. Meetinghouses and churches swelled as people began to panic. Work came to a standstill for thousands. In Connecticut, a farmer was noted to have stopped the morning chores of shoveling manure because he could no longer "discern the difference between the ground and the dung." This was much more than your usual low cloud and the duration was much too long to be an eclipse.

We are fortunate to have so many surviving diaries and writings from the time to document how such a unique day in history played out. One of the more beautiful entries is from the Braintree, Massachusetts, home of Abigail Adams. Abigail, wife to John Adams who was on a diplomatic trip to Europe, wrote down a wonderful description of the event that rivals any meteorological observer's talents.

"On fryday the 19 of May the Sun rose with a thick smoky atmosphere indicating dry weather which we had for ten days before. Soon after 8 oclock in the morning the sun shut in and it rained half an hour, after that there arose Light Luminous clouds from the north west, the wind at the south west. They gradually spread over the hemisphere till such a darkness took place as appears in a total Eclipse. By Eleven oclock candles were light up in every House, the cattle retired to the Barns, the fouls to roost and the frogs croaked. The greatest darkness was about one oclock. It was 3 before the Sky assumed its usual look . . . About 8 oclock in the Evening almost Instantainously the Heavens were covered with Egyptian Darkness, objects the nearest to you could not be discerned tho the Moon was at her full."

Just like that, a perfect timeline of events! And with some great clues for meteorological sleuthing. We'll dig into those shortly. It is clear that nearly the entire day was filled with darkness, except for a brief mid-afternoon lull.

"Dark days" are not without precedent. In the northeast alone there were instances of candles being necessary during the middle of the day in 1680, 1685, 1706, 1716, 1727, 1743, and 1762. The slash-and-burn technique of clearing land during colonial times meant fires and smoky skies were common. Many would never see or learn of a dark day to this magnitude during their lifetimes, however. In 1780 it was not possible to hop onto the internet and see videos and photos instantly confirming their existence. Dark days were localized events and you certainly would not receive word of one occurring hundreds or thousands of miles away, if ever. This was not of any help to a New Englander in the eighteenth century. What better explanation than a vengeful God? In deeply religious New England this was tailor-made for repentant end-of-times logic. Reverends thundered the arrival of Judgment Day and church attendance was noted to increase. Biblical precedents were all the rage in the 1700s.

One also must imagine the increased anxiety a colonist might be living with before the disc of the sun was blotted out. Most colonies were living near the coast on the edge of a largely unexplored continent that was full of wild animals, intimidating forests, and uncharted lands. Daily life was ruled by the unknown. To lose the sun, on the heels of a night where the moon appeared bright red, would strike some measure of fear and uncertainty into even the most unflappable.

The Connecticut Legislature was in upheaval debating whether to adjourn for the day. Among the throng calling for shutdown appeared Abraham Davenport, a member who was a little more matter-of-fact about the situation. Taking the chaos in stride, he simply requested more tools for which they could work by. "I am against adjournment. The day of judgement is either approaching, or it is not. If it is not, there is no cause for an adjournment; if it is, I choose to be found doing my duty. I wish therefore that candles may be brought."

Davenport would receive some fame for this sensibility, though in other fields many laborers did retire for the day. Some simply made off to the nearest tavern to wait it out or prepare for judgment.

Judgment would have to wait, so grabbing a pint may have been the best course of action. The first half of the following night was among the darkest ever witnessed. Samuel Tenney, a delegate from New Hampshire, described it as "as gross as ever has been observed since the Almighty fiat gave birth to light . . . A sheet of white paper held within a few inches of the eyes was equally invisible with the blackest velvet."

But this was not to be a permanent state of affairs. Later that night the wind increased and shifted, clearing out the skies. When frightened townspeople awaiting what wrath lay ahead awoke, they were relieved to see the pall was gone. A bright sun welcomed a new and less anxiety-ridden day ahead.

It may seem obvious to a twenty-first-century reader what happened to create New England's Dark Day. And it would not be factual to say everyone at the time was in a state of panicked confusion. There were several learned men who correctly guessed that a large wildfire somewhere had produced the foreboding skies. The exact location and timing of events would not be clear for more than two centuries.

If we revisit Abigail Adams diary and other accounts, we know a light rain fell during the morning. It was also observed that there was an easterly breeze to start the day. Being New England in springtime, it is an easy leap to believe that a drizzle-laden fog had moved in. Such a pattern would be extremely common for the area. At the upper levels above the ground, prevailing winds were more west-northwesterly. Those winds could easily transport wildfire smoke and ash from somewhere deeper into the heart of the continent. An intermingling of tiny water droplets in the fog bank and particulates from the wildfire remnants would make for an impressive shield to blot out sunshine, effective in bouncing the sun's rays back into space. A thick and low blanket of clouds alone can be enough to trigger streetlights. If that much is obvious, the more difficult question was location. Where exactly was this huge fire raging?

The answer came in 2007 through a dating science called dendrochronology. Researchers from the University of Missouri studied the written accounts and observations of New England's Dark Day and followed the breadcrumbs to Canada. There, a team was dispatched to the Algonquin Highlands of southern Ontario (among several other sites) to search for evidence of burn scars and to study tree rings.

Pine stumps have the uncanny ability to remain many years after the tree itself dies, leaving a road map to the past. Researchers sliced across these stumps to read the rings and find answers about atmospheric conditions hundreds of years back. According to the Director of the MU's Tree Ring Laboratory Richard Guyette, wildfires produce such intense heat that they kill living tissue inside the bark and leave an "injury" on the remaining wood when the bark falls off. Then as scientists examine the rings, they can note "charcoal formation on the outside and a resin formation on the top that creates a dark spot."

Using this technique, Guyette's team concluded that a major fire had ripped through the Algonquin Highlands in 1780 and was likely the culprit for the darkest day in New England history. The final puzzle piece was put into place!

With the advent of modern firefighting and a decrease in intentionally set fires for land-clearing purposes across North America, these events are harder to come by. But in 1950 a huge cluster of wildfires in Alberta Province, Canada, did produce a similar darkening of the skies that spread across the eastern US. Once again, there were some who used it as an opportunity to declare the end of times. So, what is old can be new again, even as scientific knowledge increases.

It may have even been inspiration for future President John F. Kennedy, who invoked the Dark Day during a 1960 campaign speech in North Carolina.

"I hope in a dark and uncertain period in our own country that we, too, may bring candles to help light our country's way."

PART THREE
Major Hurricanes in New England

Damage to US Highway 4 in Vermont after Hurricane Irene

If you ask any New England meteorologist what keeps them up at night, the answer is likely to be unanimous: a major hurricane screaming up the eastern seaboard with its eye set on our coast. When it comes to maximum destruction and impact, a hurricane is the biggest threat. Infrequent as they are, the big ones leave a mark for years after the storm passes. Extensive power outages, property damage, flooding, and redefined coastlines combine to make hurricanes a formidable foe.

The Atlantic hurricane season runs from June 1 through November 30, though storms can still form outside of this window. Early in the season, the first bubbling of swirling thunderstorms tend to take shape in the Caribbean or Gulf of Mexico. By midsummer, activity begins to spread out over the Atlantic as the ocean continues to warm. When late July rolls around, it is time for the main show to get underway—Cape Verde season. Through August and September, tropical waves move westward over Africa and eject out to sea in the area of the Cape Verde Islands. These are seedlings for hurricane development and often spend over a week traveling across the bathtub-warm waters of the tropical Atlantic. The highway they travel along between Africa and the Caribbean is called the main development region, or MDR. Many of the most powerful storms in history get their start in this hurricane nursery.

When the switch is flipped, it suddenly can feel like everything over the ocean wants to start rotating. In the northern hemisphere, hurricanes have a counterclockwise circulation and the opposite is true in the southern hemisphere. The western Pacific calls them typhoons, and Australians call them cyclones. But they're all the same in the sense that they are heat engines using warm ocean water as gasoline. The product isn't a corvette, but a steamroller of wind and water with the power to destroy.

Once a wave achieves a closed circulation with sustained winds of thirty-four knots (thirty-nine miles per hour), it becomes a tropical storm and receives a name from the National Hurricane Center. Once clocking in at sixty-four knots (seventy-four miles per hour), a hurricane is born. The naming itself is a tool used for public awareness and communication, but the names themselves are always changing over time. For hundreds of years, hurricanes in the West Indies were often named after saints depending on what day they made landfall. During World War II, it became common for many sailors and forecasters to give typhoons in the Pacific women's names. This informal convention became official in 1953 when the US began using female names for all tropical cyclones. If you've ever wondered why it seems like "female hurricanes" were on a tear in the 1950s, '60s, and '70s . . . they were. There were no male names yet!

In 1979, male names mingled with female names and we have kept an alternating mix of the two ever since. Six lists were created from A to W. There are no Q, U, X, Y, or Z storms but if there are so many that the Atlantic makes it through W, then the Greek alphabet is called up for use (Alpha, Beta, Gamma, Delta, etc.). As of 2021, the convention changed again to discontinue the Greek alphabet and instead have a supplemental list of names ready to go if the first is exhausted.

Where do all these names come from? The World Meteorological Organization, or WMO. After going through all six lists, storm names return to the beginning in year seven. The only way for a storm name to exit the rotation is for it to be retired, and the only way to achieve that is to do some serious damage. The most costly and damaging storms are removed and replaced with a new name. So there will never be another Hurricane Andrew, Hurricane Sandy, or Hurricane Katrina.

The strength of hurricanes is measured using the Saffir-Simpson scale, which rates them based on peak sustained winds. The scale ranges from 1 to 5, with a Category 5 being the top of the scale. A "Major" hurricane is not just an adjective for a powerful storm. To achieve major status, a hurricane must have sustained winds of at least 96 knots, or 111 miles per hour. This equates to a Category 3 storm on the Saffir-Simpson scale.

More often than not, the cooler waters north of the tropics weaken systems before their arrival to New England. This is especially true during June and July. By August, the sun has put in substantial work and turned the frigid ocean of March into a balmy basin able to sustain tropical systems. The most powerful and notable hurricanes in the history of the region have all arrived after mid-August, when the early-season safeguard has expired. Indeed, the "peak" of Atlantic hurricane season is mid-August through late October and this correlates with the most dangerous window of opportunity.

The Great Colonial Hurricane of 1635

"Before daylight, it pleased God to send so mighty a storm as the like was never felt in New England since the English came there nor in the memories of any of the Indians."

—REVEREND ANTHONY THACHER

IN 1620, A CONTINGENT OF BRAVE AND PIONEERING SOULS CLIMBED aboard the *Mayflower* and risked everything to reach the unencumbered shores of a new land. When they arrived, nature did not roll out the welcome mat. One must wonder if perhaps the pilgrims would have turned portside and headed toward tropical waters if they knew what they were getting into with New England weather. The winters are harsh, the seas dangerous, and the wilderness to shelter from the elements foreboding. Among the many challenges of their first years along the coast of Massachusetts Bay, none was as daunting as the Great Colonial Hurricane of 1635. Just fifteen years after landing at Plymouth Rock, the fledgling group of colonists would face perhaps the worst hurricane to ever strike the region before or since. Drownings were numerous, buildings smashed, thousands of trees crashed down, and the tide rose up to twenty feet higher than previous high-water marks.

How do we know? To be sure, there was not a vast network of weather spotters like we have now. The population of Plymouth was still under one thousand people and there were few settlers anywhere away from the coast. What would later become Connecticut consisted of a tiny settlement in Wethersfield and a small, simple fort in Old Saybrook. As sparse as this data is, there is still enough to piece together surviving accounts and match it with our current knowledge of how hurricanes operate to reconstruct this infamous storm.

The first known mention of the hurricane came from the first English settlement in Jamestown (now part of modern-day Virginia). While it was noted that on August 24 a hurricane affected the area, significant damage does not seem to have occurred. This points to the storm's center likely being out to sea over the warm waters of the Atlantic. We can also infer that it was far enough offshore to limit interaction with land since little weakening took place between Virginia and the southern coast of New England.

We might also take the leap, knowing that the end product was a powerful hurricane and the timing was late August, that it may have been a Cape Verde storm beginning as a tropical wave far across the tropical Atlantic and tracking north of the Bahamas. This would be conjecture since we do not have observations to confirm, but it would fit with climatology. Many of the strongest New England hurricanes follow this sort of life cycle. All we can say with certainty is that a major hurricane was located east of Virginia on August 24, 1635.

There is no satellite imagery or hurricane hunter reconnaissance to pick up the story from that piece of knowledge. We simply must wait for the next data point when the storm struck a spot humans happened to be living. Estimated to be of Category 4 strength at its peak and racing northeast at forty miles per hour, it slammed into eastern Long Island as a strong Category 3 hurricane and maintained that intensity as it made a second landfall near Groton, Connecticut, during the early morning hours of August 26.

Fortunately, for us, learned men were making note of this fierce change in the weather and writing down their observations as the hurricane plowed across southeastern New England. The most useful writings we have come from prominent members of society at the time, including the Governor of Massachusetts Bay Colony John Winthrop and the Governor of Plymouth Colony William Bradford.

Keeping a running diary of the event, John Winthrop noted the arrival of the hurricane in Boston right around midnight as the 26th began. The wind "came up at N.E. and blew with such violence, with abundance of rain, that it blew down many hundreds of trees, near the towns, overthrew some houses, and drove the ships from their anchors."

He also left us a great clue about the track of the storm's center by identifying a shift in the winds when the hurricane moved by. "About eight of the clock the wind came about to N.W. very strong, and it being then about high water, by nine the tide had fallen three feet."

These two passages give meteorologists breadcrumbs to identify the eye of the storm and where it was heading that morning. With the wind turning to the northwest at 8 a.m., we know that the center was passing just offshore to the east of Boston. Now there are two data points to reconstruct the track—one on the 24th near the Virginia coast but not too close as to induce weakening of the hurricane, and another just east of Boston at 8 a.m. on the 26th. Knowing those two points and the time between them also gives an accurate estimate of forward speed.

Winthrop also left us with a note about storm surge and casualties, writing that "The tide rose at Narragansett fourteen feet higher than ordinary, and drowned eight Indians flying from their wigwams." The storm surge proves to be very useful in determining strength at the time of landfall, which we'll look at in a moment.

Farther to the south, William Bradford was hunkered down at Plimoth Plantation and in awe of the raging tempest around them. He called the event "such a mighty storm of wind and rain as none living in these parts, either English or Indians, ever saw" and like Winthrop, made note of the sudden wall of intense conditions that roared in as opposed to a slow build. Like several of the highest-impact tropical cyclones to strike New England such as the 1938 hurricane, The Great Colonial Hurricane was moving fast. Indeed, his report of the damage left behind sounds eerily similar to the tales of '38. Bradford's entry on the destruction is worth repeating in its entirety:

"It blew down sundry houses and uncovered others. Divers' vessels were lost at sea and many more in extreme danger. It caused the sea to swell to the southward of this place above twenty feet right up and down, and made many of the Native Americans climb into trees for their safety. It took off the boarded roof of a house that belonged to this Plantation at Manomet, and floated it to another place, the posts still standing in the ground. And if it had continued long without the shifting of the wind, it is like it would have drowned some part of the country. It blew down many hundred thousands of trees, turning up the stronger by the roots and breaking the higher pine trees off in the middle. And the tall young oaks and walnut trees of good bigness were wound like a withe, very strange and fearful to behold. It began in the southeast and parted toward the south and east, and veered sundry ways, but the greatest force of it here was from the former quarters. It continued not (in the extremity) above five or six hours but the violence began to abate."

Oddly, both Bradford and Winthrop gave what appear to be incorrect dates for the storm, citing August 15 or 16. But Bradford's writings also include mention of a lunar eclipse two days following, which would have occurred on August 28. Combined with the Jamestown recordings, it is likely that somehow both men meant the 25 through 26 on our modern calendar.

Between the two, we now have wind direction and storm surge reports from two different locations that appear to be on either side of the eye's path. We also know it was moving fast enough to pack the worst conditions into a five- to six-hour window of time.

Colonists did not have barometers to measure the lowest pressure of the storm or anemometers to record the peak winds of the Great Colonial Hurricane, but they did have a keen sense of the tides. The coastal colonies of Massachusetts Bay were built with the tides in mind. Knowing the daily range of the ocean was critical to help plan settlements and wharves, not to mention how to sail successfully around New England's challenging shoals. Native Americans certainly shared this important survival knowledge. So when the water came up, they knew exactly how unusual it was. The high-water marks are the most reliable information we have to start figuring out all the other variables for the storm.

Over 350 years into the future, Brown University offered another clue. Their study of sediment deposits on the Rhode Island shore, published in 2001, found overwash deposits of sediment left by intense hurricanes of the past. Among them was a deposit from 1635. Their data helped corroborate the observations Bradford and Winthrop made and prove how vast the storm surge was.

There is enough to do some atmospheric detective work using modern-day technology, and that is exactly what Bill Jarvinen of the

National Hurricane Center set out to do. Jarvinen was a storm surge program leader for the NHC and an expert in modeling the damaging inundation hurricanes bring ashore. In 2006, he helped piece together what would have been unknown at the time of the storm.

Meteorologists know the relationship between a hurricane's size, strength, speed, and the surge it is able to produce. Using the observed storm surge of fourteen to twenty feet in Narragansett Bay and Buzzards Bay and the location/timing of the storm, Jarvinen ran the hurricane center's SLOSH (Sea, Lake, and Overland Surges from Hurricanes) model to figure out the rest.

The result was possibly the most severe hurricane to ever strike New England. After adjusting the model to fit the known observations, it is estimated that the Great Colonial Hurricane was a strong Category 3 hurricane with peak sustained winds of approximately 135 miles per hour. The lowest central pressure may have been 938 millibars when it made landfall across eastern Long Island and 940 millibars as it crossed southeastern New England.

After reaching the coast of Connecticut near Groton, the eye is proposed to have passed just slightly north of Providence, Rhode Island, at 7 a.m. with 130 miles per hour sustained winds, then onward to Cohasset, Massachusetts, by 8 a.m. before racing out into Massachusetts Bay with top winds of 105 miles per hour. Damage from the extensive blowdown of forests was still visible for several decades to come. Aptucxet Trading Post in Plymouth, the first site built by the pilgrims to trade with local Native Americans, was destroyed. The storm blew the roof off a building at the trading post and carried the rest from its foundation. Only the support posts were left standing.

Peak storm surge reached an incredible height of 21.7 feet at the head of Buzzards Bay. We cannot know for certain if this is the precise value, but if so it would constitute the highest surge experienced on the east coast of the US. Providence experienced a surge of fourteen feet, and Boston's high water is estimated to have been 5.2 feet above mean sea level.

These impressive marks are more than enough to allow the Great Colonial Hurricane to enter the pantheon of New England hurricanes along with storms such as the September Gale of 1815 and the Hurricane of '38. These may constitute the three strongest on record in the area, and so the return interval for these beasts appears to be on the order of 100 to 150 years.

Perhaps it was best that the region was not thickly settled during the time, considering the magnitude of damage. In today's dollars the price would easily climb well into the tens of billions of dollars. Power would be knocked out to millions, some for weeks. Even though the interior was not full of homes and businesses at the time, there was plenty to be fearful of at sea. Commerce and travel moved with the speed of your best ship in the 1600s, and so many of the harrowing stories from this particular hurricane were playing out on the Atlantic.

One such tale featured the *James*, a ship that had set out on a journey from England to Boston in June. Aboard were one hundred passengers, including Reverend Richard Mather (father to Increase Mather and grandfather to Cotton Mather, who all kept significant historical diaries over the years). The *James* had spent the past couple months crossing with the *Angel Gabriel*, which also was about to find itself in a heap of trouble.

Both reached the shoreline of the New World right as the hurricane plunged the ocean into chaos. The *James* floundered near the Isle of Shoals, due east of what is now Rye, New Hampshire. Struggling in the wind and waves, the captain lost all anchors with cables unable to survive the rough conditions. The New England Historical Society provides details about several tense hours as passengers watched the hapless ship drift closer and closer to the rocks. It would not be the first to succumb to the shoals . . . but fate had different plans for the *James*. Just before reaching the rocky islands, the storm center passed and winds reversed, blowing the ship away from destruction. The captain managed to get the hobbled vessel into Hampton Harbor and eventually to Boston after the storm.

Mather described the sense of relief on board in his diary: "oh how our hearts did then relent and melt within us! And how we burst into tears of joy amongst ourselves, in love onto our gracious God, and admirations of his kindness in granting to his poor servants such an extraordinary and miraculous deliverance."

The *Angel Gabriel*, however, did not have as happy of an ending. The ship was similar in make to the *Mayflower* but eighteen feet longer, carrying more settlers to Maine. Captain Robert Andrews managed to reach safe haven at Pemaquid Bay before the worst of the storm hit. Fortunately, the majority of the crew and passengers disembarked before nightfall on August 25. By morning, the hurricane had arrived and ripped the ship from its moorings, unceremoniously obliterating it on the rocky coast. Not everyone had spent the night ashore though, and the handful who had stayed onboard were lost.

One other ship along the shore to the south, the *Watch and Wait*, was in dire straits. This smaller boat had twenty-one aboard, including Reverend John Avery, who was making the trip from Ipswich to Marblehead to become pastor of a church there. Along for the sail were his wife, six children, and his cousin Anthony Thacher. Once coming around the tip of Cape Ann, they were met with the force of the rising storm.

Caught in the whipping wind and frenzied surf, a harrowing scene ensued with waves throwing passengers into the water. They desperately grabbed on to debris from the splintering ship to stay afloat. Anthony Thacher would later describe the tumult: "Now look with me upon my distress and consider my misery . . . my goods and provisions swimming in the seas, my friends almost drowned, and mine own poor children so untimely . . . before mine eyes drowned and ready to be swallowed up, and dashed to pieces against the rocks by the merciless waves, and myself ready to accompany them."

Wave after battering wave struck the *Watch and Wait*, and in the end only Thacher and his wife, Elizabeth, were serendipitously tossed onto an island instead of drawn down into the turbulent Atlantic. The two climbed hand over foot to the center of the island and found refuge under a cedar tree "where we sat about an hour, almost dead with cold." Thacher managed to find a backpack that had washed ashore equipped with a steel, flint, and powder horn to light a fire. For sustenance, the couple had a dead goat that had also found its way to the island. Another heartbreaking discovery, his dead son William's jacket, which had also washed up on the island and provided him with warmth to wait for rescue. Mrs. Thacher's coat was also located and helped get them through. A ship later found them and brought the couple to Marblehead. It was estimated that at least forty-six perished in the hurricane, nineteen of them aboard the *Watch and Wait.*

After the Great Colonial Hurricane, the General Court awarded the island to Anthony Thacher and it stayed in the family for eighty years. Originally dubbed "Thacher's Woe" after the tragedy, it continues to bear the Thacher Island name today. Twin lighthouses were constructed in 1771 to help prevent future loss of life along the dangerous shoreline of Cape Ann and have been beacons for over two hundred years since.

The Great September Gale of 1815

It chanced to be our washing-day,
And all our things were drying;
The storm came roaring through the lines,
And set them all a flying;
I saw the shirts and petticoats
Go riding off like witches;
I lost, ah! bitterly I wept,–
I lost my Sunday breeches!
—OLIVER WENDELL HOLMES, THE SEPTEMBER GALE

IT WAS A LONG WAIT AFTER THE CALAMITY OF 1635, AND IT IS UNLIKELY anyone complained during the respite. Colonies continued to grow and prosper, a war for independence was waged, and then yet another war to reach conclusion in early 1815. Time to rest, right? Unfortunately, there was no rest for the weary. When the battlefields quieted down, nature began to rumble. During April, Tambora's eruption in Indonesia was sowing seeds for 1816's Year Without a Summer. But of more immediate concern was a hyperactive season in the tropics. The Atlantic was set to send forth the greatest storm New England had witnessed since the days of the pilgrims, and its wrath would remain unmatched for generations to come.

The Great Gale smashed ships, tore through forests, and drowned entire families with a sudden wall of water that surged in with landfall. There are records of at least thirty-eight killed, but the true loss of life is thought to be considerably worse. Homes, businesses, and churches by the hundreds were blown apart by winds estimated to top 130 miles per hour, putting it somewhere between a Category 3 and 4 hurricane on arrival. At the epicenter, the city of Providence became an inland sea with up to a quarter of its taxable property destroyed. Sea salt was launched dozens of miles inland and was found pasted onto windows as far away as Worcester; the corrosive quality ruining foliage before it could reach autumn splendor. Crops were ruined and orchard groves splintered by the thousands.

Without question, no one alive had seen such power from nature before across southern New England. The Great Gale of 1815 swiftly delivered a gut-check that extremely intense hurricanes, though rare, did strike the region. This was a fact of life that had been forgotten by the time September of 1815 had rolled around. It is not that people were unaccustomed to coastal storms. There were often autumn wind and rainstorms, but the term "hurricane" was not yet common vernacular in the region. Such beasts were thought to be a hazard of the tropics only. A lesson in just how vulnerable humans are to this formidable force of weather was harshly learned. One could hardly blame the lack of knowledge regarding major hurricanes since there are very few that reach higher-latitude coastlines. Besides, there was no technology to track them anyway.

Had there been access to the tools of the future, the 1815 Atlantic hurricane season would have been a busy one to keep tabs on. The summer was reported to be full of destructive and calamitous storms making newspaper headlines up and down the eastern seaboard. One notably ferocious hurricane was observed in North Carolina on September 3. But it was the day of the autumnal equinox, near the typical climax of tropical activity, when the largest and most disastrous of them all would reach New England.

The Great Gale formed deep in the tropics though there are very few accounts of interaction with land or shipping vessels. Perhaps the busy season kept more ships in port than usual. But thunderstorms were blossoming and swirling near the Bahamas in mid-September. It is likely peak intensity was reached north of Grand Turk on September 20, where it is believed Category 4 strength was attained. From there, it was just a question of an escape out to sea, or a beeline straight toward the mainland US. Unfortunately, the route of choice was the latter.

New Englanders were in the dark about what was heading their way, but perhaps at least one man knew. A Massachusetts sailor who had significant experience in the West Indies was wary when extreme calm set in ahead of the storm. When asked why he was growing uneasy, the sailor said there would be a "terrific wind" before the day passed, for there was a "crackling in the air and it loomed up as he had noticed it in the tropics before an unusual wind." He would shortly be proven correct.

As is often the case, the frontline communities were on the southern coast of New England and Long Island. Landfall took place between Bellport and Eastport, New York, at approximately 7 a.m. on the morning of the 23rd. The site would be nearly identical to where the next major hurricane, the Long Island Express, would arrive nearly a century later. After blasting across Long Island, a second landfall would occur on the coast of Connecticut near Old Saybrook at 9 a.m. before continuing on a course northward through central Massachusetts and New Hampshire.

The mid-morning rush of the storm came barreling into Narragansett Bay and Buzzards Bay, where a crushing blow of water and a storm tide eleven to sixteen feet higher than normal inundated cities and towns. It was a wall of water unprecedented since the time of colonization and immediately crippled the shipping industry by ripping virtually every vessel from their moorings and smashing them apart. Widespread as the oceanic attack was, the storm's mightiest punch would land on the capital of Rhode Island.

Between 10 a.m. and noon, men drowned trying to save their property as shards of buildings and boats flew in the incredible winds tearing apart Providence. In those two short hours lives were transformed and the fortunes of the city quickly altered. By midday the wind diminished and shifted to the southwest, leaving the scene of a waterfront in ruin.

Mill shops, saltworks, ropewalks, boat builders, and blacksmiths were all swept away or heavily damaged. Meanwhile their contents became flotsam, filling the floodwaters with a variety of maritime goods. Sails, timber, cotton, and perishable items bobbed past the wrecked homes and businesses, which numbered at least five hundred.

The largest ship in the harbor, the *Ganges*, was among those being wildly tossed around in the waves and managed to wreak more havoc than most. After breaking with its mooring, the ship promptly crashed into and obliterated the Weybosset Bridge. With a new passage from the harbor to town, more boats would follow suit and scatter themselves upstream. One sloop was pushed all the way through North Providence. The *Ganges* wasn't quite done on its rampage either. Its bowsprit, swinging chaotically, managed to smash and poke holes through the third floor of the Washington Insurance Office building. A famous engraving by James Kidder, later reproduced by John Russell Bartlett, captures the chaos of this moment in downtown Providence as they were tossed and rocked in the floodwaters among the commercial district.

Editor of the *Rhode Island American*, William Goddard, would describe it as "the most sublime and tremendous elemental strife that has been witnessed for centuries by the inhabitants of this town." Once the surge retreated, there remained the unsettling picture of ships left on streets and roads barricaded by the debris of once prosperous businesses. It is difficult to watch a few fleeting moments of weather upend everything a community has worked to build. Survivors stumbled out to see water had been in places that had previously been unimaginable. A plaque would later be installed on the Old Market House in 1917 to commemorate the high-water mark that brought so much loss. Visitors can still see it today.

All along Narragansett Bay the rising tide was removing homes from their foundations and throwing around ships while men and women prayed to see the other side of the storm. In Bristol, the water came up seven feet higher than had ever been known. The wharves were lifted and carried away along with the Post Office and a number of stores. Similar destruction visited Newport, where Long Wharf was devastated and wiped clean of business. A formidable three-story building was lifted and floated away into the harbor. What had been useful in guiding boaters through stormy times, Point Judith lighthouse, was blasted apart and sent into the ocean.

The bays of New England's South Coast are one of the most dangerous places to be during a hurricane. As a hurricane approaches, wind forces water into these nooks and crannies of the coastline with great speed. Trapped with no outlet, the water is channeled into the bays and then rises into the towns that line them. While Rhode Island's coast was being forever changed, the same was occurring across the state line. Buzzards Bay of Massachusetts is another very vulnerable waterway for storm surge, and it rivaled the worst damage ushered in by the gale. In the busy port city of New Bedford, all but two ships were pushed up onto land and splintered when the ocean rapidly rose. All of the wharves were wiped away.

Floodwaters can produce some confounding visuals, and there are several reports of whole structures floating for miles, intact, to new resting places. One salt-house set sail on the tide and was found several miles away, nine feet above the typical high-water mark, standing in fine condition. Indeed, the owner located it and managed to bring it back "home." Another salt-house at Sippican (Marion) stood solemnly awaiting its destiny as the tide rose just below the rooftop. Amazingly, the flood allowed it to set sail on a course across the bay. Eventually the house's ruins were found in the woods in Wareham. Another store at Great Neck fared better—it too floated a mile away to Wareham but the goods inside somehow remained in "perfect preservation."

The ocean powered inland, burying farmer's crops in sand. Saltwater isn't friends with much of anything that lives outside the ocean itself, and the high salinity killed vegetation in many of the marshes and cedar swamps. Many wells near the coast were contaminated by the seawater and remained brackish until winter snowmelt helped to minimize the salty invader, though some did not recover completely until 1818. One group that did typically enjoy the saltwater was, as one might expect, salt producers. But the industry took a severe blow as nearly all the infrastructure to produce it was leveled.

While a massive surge piled up east of the storm's track, the point of landfall in Connecticut was dealing with widespread damage of its own. The tide in Stonington came up seventeen feet and put almost the entire acreage of the town under water. Much like the rest of the shoreline, the roaring waves left nothing but disaster. Of the twenty ships in the harbor, every one was thrown onto land or sent to the bottom. The schooner *Washington*, which had taken refuge on its sail to Newburyport, was so unceremoniously thrashed that there was not a single piece of cargo or personal belongings salvaged.

More calamity set in just west of Stonington when water surged up the Thames River between New London and Groton. No one could recall such inundation. Even eighteen miles upstream in Norwich, the river swelled high enough to sweep stores off their wharves and into the Thames. While not quite as severe, an impressive surge down Long Island Sound damaged bridges and flooded homes down the coast to New Haven.

To the east on Cape Cod, the effects of the gale were significant but much less severe once east of Buzzards Bay. This makes the case that the storm was still very tropical in composition, with the core of its power still consolidated near the eye instead of the broader wind field of an extratropical cyclone. There was also less coastal damage along the north shore of Massachusetts, though the howling winds were full throated enough to wipe out trees and thrash property.

Though not taking the brunt, it was still a storm unlike any other for thousands of Bostonians. At daybreak the previous day, the weather in New England had not signaled tropical fury on the horizon. A cool and heavy rain began to fall though the barometer was not yet falling in tandem. For nearly twenty-four hours it poured as northeast winds blew down from the Gulf of Maine. Making note of the gloomy conditions was a Harvard professor of mathematics and natural philosophy by the name of John Farrar. From his perch in Cambridge, Massachusetts, Farrar would take note of the deteriorating conditions and chronicle the storm in a publication four years later. Not only would the observations provide a detailed story of the gale's progress, but a hypothesis that we now know to be true—a hurricane is a storm with a counterclockwise circulation (northern hemisphere).

Farrar would describe how the moderate and cool winds of Friday became something wholly different on Saturday morning, the 23rd of September. At first, it was a shift of the wind to the east and a freshening of its speed. The gusts came harder and faster, turning more southerly as the morning wore on. By 10 a.m., the core of strongest winds, which wrap around a hurricane's eye, began to ravage the city of Boston. For an hour and a half, the height of the storm destroyed property and "excited alarm."

Chimneys and trees were blown over and shingles sent flying like missiles across the city. The Charles River "raged and foamed like the sea in a storm, and the spray was raised to a height of sixty or one hundred feet in the form of thin white clouds." Fires were unable to be maintained inside homes, so great was the force blowing across the tops of chimneys.

The intense hurricane blew down many of the older and weaker buildings and churches, and laid waste to the leafy landscape in particular. Strong enough were the gusts that they bent the steeple of the Old South Church. Farrar described the scene as desolation "which has never been witnessed before to such an extent in this country." Roads were impassable and it was estimated that five thousand fruit trees were destroyed in Dorchester alone on the southern side of the city. Fortunately, the height of the onshore wind did not coincide with high tide in Boston and prevented a catastrophic surge from reaching full potential, though destruction of ships and wharves was considerable. At least sixty vessels were sunk or wrecked in Boston Harbor and a schooner would be found sitting on Main Street in Cambridgeport, shoved inland by the excited ocean.

Throughout the countryside west of the coast, a white "frost" of salt could be found on plants, trees, and homes. A curse on living foliage, the briny coating turned leaves brown, crisping and curling them long before their more typical seasonal demise. Grapes harvested later in the season tasted like salt. The brackish spray made its way into rivers and streams as far inland as Sterling, Massachusetts, which sits roughly forty miles from the ocean. A briny taste in the water lingered for days. In some localities where freshwater was scarcer, it was said that some springs did not completely rid themselves of the salt for years. It wasn't just particles in the air moving to unfamiliar territory, but wildlife as well. Seabirds found themselves in the middle of New England forests, including gulls and petrels.

Before this September day, it was widely assumed that a hurricane was a wall of wind that rushed straight forward from the tropics, knocking down everything in its path before blowing itself out. Farrar's keen observations would provide evidence that was not the case, though it would not be fully accepted for a couple more decades. The professor would compare his records of powerful veering winds to the southeast to the northwesterlies in other locations across New England, New York City, and Quebec. His conclusion was that "it appears to have been a moving vortex." It was a powerful realization that helped push forward understanding of how tropical meteorology works.

The Great New England Hurricane of 1938

"Our house—ours for 25 years—all our possessions, just gone. My God, it was something devastating and unreal, like the beginning of the world or the end of it."

—Katharine Hepburn

Imagine waking up on a perfect sunny and 75°F September day. You are whistling your way to work while enjoying the final warm moments of summer. Or you are a schoolchild gleefully skipping to class, still reveling in the start of a new year. Perhaps you are a fisherman setting a course toward the red sunrise, thinking about how successful the day's catch may be. Then, within a few hours, the biggest weather calamity in over a century rips apart your world and sets a benchmark all other hurricanes will be measured against. No matter which of these characters you were, you had no idea such an event was possible when you first looked out the window. This was life on September 21, 1938.

If you are ranking all-time New England hurricanes, '38 starts and ends the discussion. Second on the list is quite a ways down. Over the course of just five hours that day, forests were decimated. Storm surge inundated and annihilated coastal communities. Thousands of homes, cars, and boats were destroyed. Hundreds of lives were lost. No other storm has racked up a dizzying array of sobering statistics in such a short period of time.

In modern times, it goes beyond comprehension that such a monumental event could go relatively unnoticed until landfall. We would have chatter about favorable hurricane conditions weeks in advance. Up next, outlooks and National Hurricane Center cones showing the path ahead. Watches and warnings posted. Media coverage would start so early you might even be tired of hearing about the storm by the time it arrived. For those in 1938, there were no such luxuries. Satellite coverage, internet service, cable television, and smartphones would not be of any help. The only real notice given to New Englanders and Long Islanders on the morning of the storm were the brilliant sunrise as cirrus clouds fanned out ahead of the hurricane and the barometer ominously dropping like a rock.

The lack of technology and a path that avoided contact with land until the last minute made '38 a difficult storm to track or warn. It was a classic Cape Verde hurricane, forming just off the coast of Africa on September 9 and taking a nearly two-week journey across the Atlantic. Since it was tough to observe storms, this meant it was up to ships at sea and outposts in the Caribbean Islands to relay what they were seeing and raise the alarm. There were a lot of blind spots and educated guessing was the only way to piece things together.

Even with this rudimentary warning system, information was flowing to the Weather Bureau Office in Jacksonville, Florida. On Sunday the 18th, a Brazilian merchant ship, the *S.S. Alegrete,* spotted the hurricane about three hundred miles north of Puerto Rico. It is believed peak intensity, likely Category 5 strength, was achieved around this time as it approached the Bahamas. Forecasters were aware that a large hurricane was out there and believed that it was on its way to Florida.

On the morning of September 19, warnings were issued for the entire Atlantic coast from Key West to Jacksonville. By the next day it seemed they may not be necessary. It was becoming clear the severe hurricane had started to make a turn to the north. The center was located about four hundred miles east-southeast of Jacksonville and storm warnings were extended up the eastern seaboard to Atlantic City, New Jersey.

Word was out, but on the 21st both forecasts and conditions deteriorated rapidly. Few mariners were out on the ocean anymore, thanks to the warnings from Jacksonville's WBO. Unfortunately, this is where the warnings ended. At 7:30 a.m., the last advisory from Jacksonville's WBO was issued and control was handed over to Washington, D.C.'s office. Another extension of storm warnings was made up the coast from Block

Island, Rhode Island, to Eastport, Maine, at 9 a.m. Forecasters only discussed a tropical storm, severely underestimating the strength of a much larger and stronger hurricane. They had been misled by the lack of critical observations.

Just two and a half hours later, at 11:30 a.m., the next advisory removed any mention of a tropical storm or hurricane. Now it was simply a gale that would diminish overnight. At the same time this was sent out, the worst hurricane to ever hit New England was only 175 miles south of Fire Island, New York.

And then the final advisory issued at 2 p.m., reporting that the center was seventy-five miles east-southeast of Atlantic City. Washington conceded that a storm would cross Long Island, but by this point gusts over seventy miles per hour had already reached there and coastal Connecticut. Children were let out of school early amidst falling trees, boardwalks were being shredded. And then landfall came roughly half an hour later. The New York WBO never even received any information about the strength and location of the storm until it hit.

The hurricane made an initial landfall in Bellport, New York, between 2:10 and 2:40 p.m. with howling winds sustained at 120 miles per hour, strong enough for "Major" Category 3 status. The lowest pressure was measured here, bottoming out at an ear-popping 27.94 inches. With very low pressure came a fifteen-foot storm surge bringing devastating impacts across Long Island.

A rapid rush of ocean forcefully removed several homes from their plots and pushed them clear across Long Island Sound, where some eventually ended up on the coast of Connecticut. Several died trying to hold on to their rooftops like rafts to make the trip. In Westhampton Beach, 153 out of 179 homes were instantly destroyed. Ten new inlets were carved through the sand, like Shinnecock Inlet that still remains today. In a matter of an hour or two, twenty thousand miles of telephone and power lines came down. After crossing the Sound, it remained a Category 3 hurricane and made a second landfall near 4 p.m. in Milford, Connecticut. Even though it may have been undergoing extratropical transition at this time, the eye was still coherent and was observed in New Haven, Connecticut, just to the east of Milford.

There are many different ways a tropical cyclone's toll can be measured, but the most deadly and destructive element is often storm surge. What appeared to onlookers as a fog bank was in fact a ten- to twenty-foot wall of water traveling fifty miles per hour, packing far more power than dunes or dwellings could slow down. So much raw energy was stirred up by the cyclone that a seismograph four thousand miles away in Alaska detected vibrations. Adding on to the tremendous surge was a landfall that timed out precisely with high tide and during a full moon, the worst possible moment. Even in New York Harbor, west of the center, the water level rose seven feet in only thirty minutes.

The hammer fell hardest on Rhode Island's coast where surge was funneled up into Narragansett Bay. People who had been casually going about a pleasant beach day in coastal Westerly, Rhode Island, were caught by the suddenness of a summer thunderstorm. In a matter of minutes the wind roared, trees started coming down, and a lightning-quick surge swallowed up anyone too close to the ocean.

In another instance of tragic timing, the storm struck right as children were getting out of school. The saddest tale is of a bus in Jamestown, swept off the Mackerel Beach Causeway by the tide. The bus driver, Norman Caswell, attempted to save the students inside, but a huge wave crashed through the rescue effort and drowned seven children. A father who was on the causeway had to watch four of his own drown.

As gut-wrenching as this scene is, it was one of many similar ones taking place all over the area. Ten women in Misquamicut who were at a church social all died when the cottage they were in was shredded by waves. Watch Hill and Matunuck were inundated with dozens of people claimed by the sea. Over one hundred would perish in Westerly and neighboring Charlestown, where 160 of the 200 homes were destroyed. The Westerly High School had to be turned into a temporary morgue. The toll was so great there was a shortage of embalming fluid. In all, 380 of the nearly 700 total souls lost were residents of Rhode Island.

Wind-driven water continued its journey northward into Narragansett Bay, where it submerged Providence, Rhode Island, under a storm tide of nearly twenty feet. Sustained winds were clocked at one hundred miles per hour with a gust to 125 miles per hour in the city. It was the end of the workday but people went to higher floors of office buildings for shelter to wait for their chance of escape. Newport's Ocean Drive was torn apart, and of a hundred ships that frequently fished the waters between Point Judith and New London, Connecticut, only three remained in the storm's wake.

As if the scene wasn't end-of-times enough, the storm did not even have newspaper headlines to itself. The *Pawtucket Times* featured a front page "300 gale deaths forecast, Europe on brink of war." And on top of it all, many were still reeling from the Great Depression.

Along the coast to Connecticut the scene was no different as record tides pushed into the cities of New London, Stamford, and Bridgeport to name a few. The unprotected southeast corner, open to wave action at the far end of the Sound, absorbed the brunt. A one-thousand-ton lighthouse tender, *Tulip*, was chucked onshore in New London. While it rested on top of the railroad tracks behind the Customs House, making for a wild juxtaposition, the ship *Marsala* crashed into a warehouse and started a fire that destroyed a quarter-mile section of the business district. All the city's docks were wrecked. Water did not stop at the coast as it pushed into estuaries and rivers, including the Thames, which filled the city of Norwich with eleven feet of water.

On the South Coast of Massachusetts, another vulnerable area quickly filled with water. Buzzards Bay forms a bottleneck where there are no outlets for the ocean once the wind forces it in. The city of New Bedford immediately was inundated with eight feet of water. The famed South Coast fishing fleet was devastated in short order as two-thirds of the boats moored in New Bedford's harbor were sunk. Though this was the tip of the iceberg when it came to the staggering toll exacted on mariners. Across New England, 2,605 boats were destroyed, with an additional 3,369 damaged.

Across Buzzards Bay to Falmouth and Bourne on Cape Cod, it was the same story. Storm surge ripped up oceanfront property and flooded towns. Along the western edge of Martha's Vineyard, the Menemsha waterfront was obliterated. A passage in the *Vineyard Gazette* several days later shed light on the shock and surprise felt by many on the island, even the Coast Guardsmen.

"The circumstance which caught even the oldest and most experienced fisherman unaware was the direction of the wind, which was southeast, becoming south. Wind from this direction has never been known to reach a high velocity on the Vineyard or to cause serious damage."

One of the most difficult aspects of forecasting or wrapping your head around the most extreme events is that few ever actually see one. No one alive in 1938 had ever been through such a storm. The last time anything of equal intensity had been the Great September Gale of 1815. Before that, the Great Colonial Hurricane of 1635. So there were no past experiences with which to set expectations or prepare. A lot of people simply did not even know what a hurricane was, let alone something this vicious.

For inland areas, the scourge of the storm transitioned from towering waves to flooding rain and intense wind. It had already been a wet summer, and a wet week leading up to the storm's arrival. Even though no one expected a hurricane to strike, they were well aware of the flooding risk from more incoming rain. That part of the forecast more than verified, and just two years after the epic flooding of 1936.

The Connecticut River Valley was ground zero for rising rivers. Between the hurricane and rainfall over the preceding few days, ten to seventeen inches fell on the basin. Sections of the New York, New Haven, and Hartford railroad lines, already a mess from fallen trees and wind, were washed away. The swollen Connecticut River in Hartford, after just reaching a record level of 37.6 feet in 1936, rose to 35.4 feet. To this day, they represent the two highest water marks ever recorded there. No other flood has come within 4.5 feet (thanks in part to more flood control measures in the future).

By now, the surge had ravaged New England's South Coast and floodwaters had swamped communities across western New England. To the east, rain was far less but wind became the headline. The longest continuously operated climate site in the US is Blue Hill Observatory, atop a hill just south of Boston. On September 21, it clocked winds that haven't been seen since. Standing 635 feet above sea level, observers used a nephoscope to calculate that the storm was now moving at a forward speed of sixty miles per hour. In other words, air passing by Providence, Rhode Island, was in Milton, Massachusetts, three minutes later.

Roaring blasts of wind peaked at 186 miles per hour. If you are wondering how such tremendous winds were measured, that peak gust is actually an estimate. Readings by several observers established a top five-minute average wind speed of 121 miles per hour from the south, from which the 186-mile-per-hour gust was calculated with an uncertainty of thirty to forty miles per hour. Regardless of the technical aspects, the effect on structures and the land all around was tremendous.

The total tally on loss of property features 8,900 homes destroyed, 25,000 homes damaged, 26,000 automobiles wrecked, and whole fleets of ships and marinas laid to waste. Even in Depression-era dollars, it was an estimated price tag of over $300 million. In today's dollars? Billions of dollars. And if the same storm were to hit in a post-2020 world, the costs would surely soar into the tens of billions.

The material losses were profound, but in the end rebuildable and replaceable. New England's landscape, however, was another matter. Forests were thrown down like matchsticks and were forever changed by just a few hours of extreme winds brought up from the tropics.

A tree in this region of the world tries to prepare itself for life. That life is full of all sorts of winds from sea breezes to nor'easters and arctic blasts, but the most common direction from which to get strong gusts is from the northwest to northeast direction. So roots grow and anchor to brace against those prevailing winds.

To get severe wind from the southeast is much more rare and it puts trees in a precarious position. The '38 hurricane took full advantage of this situation. In one afternoon, 5% of all forest in New England came down.

How much is 5%? If you've ever gone on a hike up in New Hampshire among the White Mountains, you feel a vastness around you in all directions. Well during this single hurricane, as many trees came down as are in the entire White Mountain National forest. A total of 2.6 billion board feet are estimated to have been felled by the extreme southeast winds across 904 different towns. To fathom what that would look like, picture four hundred thousand fully loaded logging trucks rolling by.

They came down near the point of landfall, but also well inland. Central Massachusetts, Vermont, and New Hampshire were ravaged by strong gusts still in excess of one hundred miles per hour, forever changing the appearance of the countryside and towns. Earlier generations had taken to clear-cutting of land for farming, but when farms were abandoned white pine had quickly filled in. They became '38's biggest woodland victim. White pine would account for 90% of the salvaged wood in the years after the storm. A tall and fast-growing species of tree, whole stands were pushed right over when the hurricane barreled through.

You can partially thank their demise for the currently splendid look of New England's forests. Without the towering pines to block out daylight, hardwoods immediately took advantage. Maples and birch trees made themselves cozy during the respite from competition that previously held them back. If you are a fan of deciduous trees over conifers, '38 helped your team succeed. We reverted to the forest as it was before colonial farming.

That is nice now, but after the storm there was a gigantic mess that threatened to crash the lumber industry thanks to an influx of supply. It was so out of hand that a new federal agency was created after the storm to try to deal with all the lumber. The Northeast Timber Salvage Administration (NETSA) put thousands of men to work removing, chopping, sawing, and milling logs throughout the region. In all, six hundred thousand acres would be cleaned up, an area slightly smaller than the state of Rhode Island. It would take a decade for paper mills to process all the wood.

And there was a rush. Not only did everyone have the herculean task of rebuilding and picking up, they had to also worry about a different potential disaster to follow—wildfires. A landscape littered with dead and drying wood was ripe for fire. The government tasked several agencies to work on mitigating the risk. Chief of the division of forest protection for the US Forest Service, Earl Peirce, noted in a later report that 4,876,519 man-days of labor were spent on fireproofing activities. In addition to cleanup, lookout towers were repaired and water holes were created for firefighting efforts. Thankfully, a massive, region-wide wildfire never developed.

If we could point to one main reason that made '38 a disaster of biblical proportions, it comes down to speed. Hurricanes do not blast forward at speeds you'd set cruise control for on the interstate. Most will travel at the leisurely pace of ten to twenty miles per hour. Sometimes if the steering currents are particularly strong, up to thirty miles per hour. How fast was '38 going? At landfall, the forward speed was fifty miles per hour.

The pattern in place made for a perfect "slingshot" that propelled the storm at a highly unusual rate of speed. A stalled cold front on the east coast was helping to guide and accelerate it northward, while a large ridge of high pressure to the east was also pushing it along and blocking its escape route. Most hurricanes re-curve and head out to sea once they're off the coast of North Carolina, but the combination of these two elements would push '38 due north and even bend the storm inland once making landfall.

There are several reasons why the speed was so crucial in making '38 infamous. For starters, you can see how quickly the situation spiraled out of control for forecasters. Racing along gave fewer opportunities to get eyes on the storm or any reports from ships. The center was also able to rocket over the Gulf Stream, maintaining a warm water source that kept it from weakening on approach. Very few hurricanes ever make it to the cooler northern waters as a "Major" Category 3 or stronger. The '38 hurricane was the first in more than one hundred years.

Of course, the fast speed also meant no warning for unsuspecting people onshore. There were no early signals telling people it might be time to hunker down. It blew in so fast and furious there was nothing to do but cling to the nearest safe structure and hope for the best.

And finally, there is relativity to consider. Winds around a storm flow counterclockwise in the northern hemisphere, which means you will find the worst storm surge, winds, and damage to the right of center (relative to its motion). With '38, the extreme forward motion helped to produce the highest possible impacts in that region, which included most of New England and Long Island. Because of this relativity, wind, in particular, was greatly increased on the right side of the track. You can essentially take the winds of 115 miles per hour, add fifty miles per hour due to its forward motion, and get winds of 165 miles per hour (the sort of which were seen on Blue Hill) on the right side of the track. Conversely, you would subtract fifty miles per hour from the winds left of center. In New York State the wind gusts were dramatically lower than in New England, as was the damage.

After all this, you may be wondering why we just call it "The '38 Hurricane" or "Long Island Express." We're used to actual names for tropical systems now. It wasn't until 1950 that storm names officially were given by the National Hurricane Center, and not until 1979 that the list as we know it (alternating male and female names) began.

Whether a hurricane to rival '38 will strike again is a matter of time. Whatever it is named when it gets to New England, we can be assured that the resulting aftermath would be unlike anyone currently residing in the region has ever seen. Power outages for weeks, if not months in some areas. Tens upon tens of billions of dollars in damage. Neighborhoods forever changed. If there is a single hypothetical weather event that keeps meteorologists up at night, this scenario is it.

Hurricanes Carol and Edna, 1954

"Looking SW out of 2nd floor of the Kingman Marina lighthouse window, a wall of water resembling a drive-in movie screen stretched from Scraggy Neck across Buzzards Bay to Mattapoisett. From the time it took to run down the stairs and take about 7 steps on the pavement the water had reached above my ankles. Another 4–5 steps and the water was over my head."

—THOMAS KINGMAN, KINGMAN MARINA

ANYONE WITH A KEEN INTEREST IN WEATHER AND ITS HISTORY HAS probably noticed that certain periods tend to have a "claim to fame," depending on how the patterns of our oceans and atmosphere set up. The 1930s are famous for the "Dust Bowl" with widespread heat and drought. The 1940s brought a barrage of hurricanes to Florida. The 1960s into the '70s are known for some pretty fierce winters. And the 2010s for record heat and heavy precipitation as the climate warms up. These are just a few of our most recent weather epochs.

So far as tropical weather goes in New England, the 1950s take center stage. No other stretch in modern times comes close to this run of high-impact hurricanes and tropical storms. Ahead in the next chapter we'll look at the disastrous flooding from Connie and Diane in '55, but before them came one of the strongest hurricanes to ever strike the region. In fact, it would become the first US hurricane to have its name retired. A little over one week later, one of equal strength would follow.

Carol, as the first of these beasts became to be known, came into existence near the Bahamas on August 25, 1955. We have a saying in New England meteorology that anything tropical north of the Bahamas is our business and needs to be taken seriously. And so it was with Carol.

The tropical wave quickly developed a closed circulation on the 26th and intensified to a tropical storm. As hurricane hunters flew overhead they could already note a circular eye about twenty nautical miles wide. By the 27th Carol was a Category 2 hurricane with one-hundred-mile-per-hour sustained winds.

At this point in its life cycle the storm was not in a hurry to move and drifted sluggishly off the coastline of Georgia and South Carolina from August 27 through the early morning hours of August 30. Then, finally feeling the effects of a trough in the jet stream approaching the eastern US, it began to pick up speed and purpose. Fortunately for North Carolina this kick kept the storm slightly offshore and spared them the worst. Still, a peak gust of seventy-eight miles per hour was notched at the Weather Bureau station in Hatteras, North Carolina, and significant flooding was felt along the immediate coastline. Carol's closet pass to the Outer Banks came on the evening of August 30.

Now that the intense storm had gotten a jolt from atmospheric steering currents, it rocketed northward toward Long Island and New England with a forward speed of over forty miles per hour. Just twelve hours after making an approach to the barrier beaches of North Carolina, Carol made landfall on another spit of sand protruding out into the Atlantic. At approximately 9 a.m. on the morning of August 31, the center crossed Long Island just to the east of Westhampton as a Category 3 hurricane. Ninety minutes later the last major hurricane to make landfall in New England (as of spring 2021) crossed Long Island Sound and reached the mouth of the Connecticut River between Old Saybrook and Groton.

By the time most hurricanes reach our shores, they often do not look much like a classic hurricane anymore. They're getting shredded by upper-level winds and feeling the effects of colder ocean temperatures, starting their transition to an extratropical cyclone. But not Carol. One of the more remarkable photos captured during Carol's landfall is from the Old Griswold Hotel in Groton, Connecticut, where beautiful blue skies are seen overhead. They were inside the eye of the storm, where a minimum pressure of 957 millibars was also observed. Carol had a true eye when it arrived, something rarely witnessed in New England. To this day it is thought of as one of the most purely tropical of all hurricanes to strike so far north in the Atlantic.

Wind and waves were the most extreme witnessed since the infamous Hurricane of '38, and in many of the same areas. The brunt of Carol's wrath was focused on the right side of the track across southeastern Connecticut, Rhode Island, eastern Massachusetts, and Maine. Widespread sustained winds of eighty to one hundred miles per hour were notched across this region blowing down trees and power lines. The tiny but beautiful Rhode Island outpost of Block Island recorded its highest wind speed of record at 135 miles per hour.

There is a remarkable set of observations from the Blue Hill Observatory outside of Boston as Carol raced northward. At 10:23 a.m., the observer notes several gusts over one hundred miles per hour and stones blowing off the WGBH roof. At 10:45 a.m., diminishing rain and brightening skies. And then at 11:09 a.m., the edge of the eye! The observation reads "Clearing, hole in So with blue sky and Ci above. No rain, very brief look through trap door." In other words, patches of blue sky seen to the south with some wispy cirrus clouds as the center of the storm moved just west of the observatory's Milton location. From 10:16 a.m. until 11:43 a.m. there were numerous gusts in excess of one hundred miles per hour with a peak wind of 125 miles per hour.

Carol's fast movement brought the storm center into New Hampshire by 1:30 p.m. while it quickly lost tropical characteristics. The wind damage left behind was severe, knocking out power and leaving the smell of shredded trees and sap in its wake. One of the more iconic losses was to the historic Old North Church in Boston, where the midnight ride of Paul Revere had begun back in 1775. Onlookers watched as the steeple swayed in the extreme winds before it eventually succumbed to the force of the storm. Fortunately, there was enough warning to evacuate the neighborhood and no one was hurt when it came crashing down on Hull Street. The original steeple dating to 1740, this was the second time the weather had conspired to rip it off (the other during the Great Gale of Boston in 1804). Nature has a sense of humor though, as the original weather vane atop the structure survived both falls and still exists today.

Down the road in Allston, a large piece of WBZ-TV's transmission tower was toppled and came crashing down, destroying a portion of the building. Housing both television and WBZ Radio 103 operations, it was a sudden loss for the flow of information to the public. The intrepid crew reacted quickly and WBZ radio was only off the air for three minutes.

Late summer also being peak harvest season, the wind's devastation to agriculture was immense. An estimated 40% of the apple, corn, peach, and tomato crops from eastern Connecticut to eastern Massachusetts at a loss of over $15 million alone.

Even though Carol was weakening as it went, New Hampshire's seacoast northward to Maine was not spared. At the Hampton, New Hampshire, airport several small planes were ripped from their anchoring and destroyed. Several dozen boats were damaged or destroyed in Rye Harbor. There were evacuations before the worst struck, as many seacoast residents retreated inland to avoid the fury of the wind and ocean. The Phillips Exeter Academy opened up as a storm shelter for six hundred people to ride it out. Gusts up to eighty miles per hour were recorded as far north as Augusta, Maine.

That was just the wind side of the story, but like many other fast-moving and potent storms, Carol brought a sudden and destructive storm surge right into the same communities devastated by '38. In so many instances throughout weather history, the bottom line of how damaging a storm is comes down to timing. For Carol, the timing was terrible. The surge arrived just after high tide, adding an extra boost to water levels. Narragansett Bay and New Bedford Harbor fared the worst with a peak storm surge of 14.4 feet crushing low-lying areas. Shoreline just east of Carol's center, from New London to Cape Cod, was inundated. The ocean once again surged right up into the city of Providence, covering it in a salty blanket of water twelve feet deep. The popular coastal town of Westerly lost two hundred homes to the sea. In neighboring Connecticut, waterfront neighborhoods in Groton, Mystic, and New London were identifiable only by the tops of buildings sticking up above the waves. The New London storm tide reached 9.6 feet mean lower low water (MLLW), a foot shy of the 1938 record (10.6 feet).

There was, at least, a piece of good news. Forecasting was certainly not perfect (and still isn't), as we'll see with Diane in the next chapter. But for Carol, warning was given and people did have some time to prepare and react. Massachusetts state officials called for an evacuation of Cape Cod and twenty thousand people listened, which undoubtedly saved lives.

All told, the damage left behind was staggering. Carol destroyed nearly 4,000 homes, 3,500 cars, and more than 3,000 boats. The price tag came to the tune of approximately $460 million, or roughly $4.5 billion in 2020 dollars adjusted for inflation. At the time, it was the costliest hurricane ever to strike the US. There were sixty-five people killed by the hurricane. Those picking up did so without power or phone service for a period ranging from a couple days to several weeks. About one-third of New England wound up losing power with as much as 95% of all phone communication disrupted.

Having to deal with one of the worst hurricanes of the century is enough for one summer. Sadly it wasn't over by a long shot. The 1954 hurricane season was wildly active and destructive for the eastern US with storms targeting the coastline like buckshot starting with Carol in August and ending with Hazel in October. Less than two weeks after Carol's landfall, another powerful hurricane was on the way for a remarkable twelve-day stretch in New England.

Edna must have looked up to its big sister, because it would be difficult to draw up a more identical storm track if you tried. On September 5 the initial tropical storm formed north of Hispaniola and rapidly intensified into a hurricane as it curved north of the Bahamas. The first hurricane hunter flight sent to investigate the storm on September 7 noted a central pressure of 1001 millibars and an eye with the width of twenty-eight nautical miles. From there, it is a little bit of a mystery as to what Edna was doing for a couple of days. There were only two reconnaissance flights September 8 through 10 and no reports of hurricane-force winds from coastal stations or ships at sea. Then, during the evening of September 10, a hurricane hunter flight found central pressure was down to an impressive 946 millibars. Another powerful hurricane was lurking east of North Carolina's coast. Once again, it would spare the Outer Banks the worst and make a beeline toward New England.

Senses were heighted along the eastern seaboard after Carol's rampage, and so there were numerous evacuations from the New York City area eastward to Massachusetts. No one was going to mess around after what they'd just seen. Businesses were shuttered and homes boarded up. The Navy and Coast Guard evacuated ships and planes anywhere near Edna's potential track. At Fort Devens in Massachusetts, General Gjelsteen sent forty trucks full of blankets, cots, mattresses, and generators to the South Coast. At the very least, this preparation is credited with the lower death toll. A total of twenty died along Edna's path.

Cape Cod and the Islands of Martha's Vineyard and Nantucket did not take a direct strike from Carol, but it hadn't been smooth sailing. Winds gusted upward of eighty miles per hour, property damage was significant, and coastal zones had been flooded. Of this one-two punch of tropical trouble, here Edna would strike harder. Like its predecessor, Edna was a rapidly moving storm with a forward speed of forty-five miles per hour, plowing into Martha's Vineyard and making landfall as a strong Category 2 hurricane just after midday on September 11. In Chappaquiddick, blue sky and a rainbow inside the eye was captured on film. It is truly astounding to think that not one, but two eyes were documented on the southern New England coastline within two weeks of each other. Many in the area will go their entire lives without ever seeing the clear eye of a hurricane! A second landfall minutes later would occur just west of Hyannis, with a third coming five hours later near Eastport, Maine.

New Englanders being matter of fact about things, a *Vineyard Gazette* interview with Menemsha's Everett Poole sums up the general mix of acceptance and disbelief after the two hurricanes: "I was here for Edna, which actually blew harder than Carol. But everything was already wrecked, so it did not do a hell of a lot of damage."

Poole's assessment aside, there was plenty of destructive wind and flooding to go around. An anemometer at the airport on Martha's Vineyard clocked a peak gust of 120 miles per hour on the backside of the storm (out of the northwest). Fresh off the strongest gust ever recorded on Block Island, winds roared over the century mark again, maxing out at 110 miles per hour. Hyannis reached speeds of one hundred miles per hour while Boston notched a top gust of 101 miles per hour. Nearly all of Cape Cod and the Islands lost power.

Perhaps only one thing was working in Cape Cod's favor, and that was the timing. Unlike Carol, Edna hit at low tide. So, in the end it could have been much worse. Even still a six-foot surge pushed into the South Coast and over already battered beaches worn down by Carol. Many boats that were left standing from the destruction twelve days earlier were destroyed or swept out to sea.

Technology at the time was growing by leaps and bounds as scientists scurried to grab whatever data they could to better understand storms. In this sense, Edna offered a rare opportunity. The Massachusetts Institute of Technology Lincoln Laboratory site in South Truro (outer Cape Cod) was able to capture some of the best eye imagery of the time. The subsequent study of that imagery helped to understand how air rises in a tropical system and then descends in the eye, clearing out the center of the storm.

After battering southeastern Massachusetts, Edna continued to move northward where some of the worst damage was yet to come. Even with the center largely offshore, extratropical transition in the Gulf of Maine meant that Edna was becoming larger with widespread gale-force winds and torrential rainfall. So fierce were the wind gusts and flooding rains that Edna was destined to become the costliest hurricane in Maine's history with damage totaling $25 million.

Areas near the coast were absolutely drenched with five to eight inches of rain and rivers rapidly pushed over their banks. The city of Portland picked up 7.49 inches of rainfall, much of it falling in six hours during September 11. Adding insult to the rain were wild winds that gusted to seventy-four miles per hour. With a soaked ground, it was more than enough to push over hundreds upon hundreds of trees. Since Carol had passed to the west and Edna was passing to the east, the tremendous number of downed trees crisscrossed each other on the forest floor, being felled by opposite wind directions.

Rivers and streams exceeded levels seen during Carol and, for that matter, the spring snowmelt season. The Kennebec and Androscoggin were among those with the most destructive rises, washing out roads, rail lines, and bridges as the power of water rushed downstream.

In the town of Unity, located between Augusta and Bangor, a family of ten hung on for dear life to the roof of a car for seven hours while floodwaters raged around them. A human chain of people rushing to help attempted a daring rescue, and it was almost successful. Tragically an eight-year-old child drowned, as well as a rescuer by the name of Alton McCormick. The rest made it back to dry land safely.

Many historic storms spur on new research and development for the meteorological community, and the hyperactive 1954 season certainly put a lot of attention on the need for better tropical forecasting. Since Carol had been such a sensation, there was immense attention directed at Edna. The esteemed Edward R. Murrow and a CBS television crew flew along with hurricane hunters on a reconnaissance flight. Part of the story that resulted featured an interview with the Weather Bureau's Bob Simpson, who passionately made the case for more hurricane research. Plus, Murrow ended the piece with a famous poetic flourish: "In the eye of a hurricane, you learn things other than of a scientific nature. You feel the puniness of man and his works. If a true definition of humility is ever written, it might well be written in the eye of a hurricane." Lo and behold, Congress allocated funds for the National Hurricane Research Project to further study the storms via aircraft.

One would think that two storms of such power, arriving so close together, is about as bad as it could get. New England was hurting, but it could have been even worse. The astute reader may have noticed that we went from Carol to Edna in a very short amount of time, so where was Dolly? Believe it or not, there was another powerful hurricane that snuck in between Carol and Edna, and it wasn't far away. Dolly tracked just three hundred miles east of Edna's track, moving off the North Carolina coast and then just east of Nova Scotia. It is mind-boggling to think of what would have been produced if three straight hurricanes hit within twelve days. Thankfully, it did not come to pass.

PART FOUR
The Power of Water— Major Floods

Connecticut River Flood

IS THERE A FLOOD IF NOTHING IS IN THE WAY? IT IS INTERESTING TO think of life before we put all sorts of things in the path of the waterways across New England. The more mobile periods of human existence, hundreds of years ago, were far more equipped to handle floods on major rivers. You could plan on spring flooding and expect to vacate the floodplains during that time. When the water receded, time to move in and farm. There were still plenty of opportunities to be caught off guard but some of the larger events could be avoided. Then European settlers started to arrive (by water, of course) and the game began to change.

Since then, the expanding human footprint on land has been constant. Growing communities and the buildings that support them can't pick up and head to higher ground for a couple months. The proliferation of dams has proven to be a successful method to control the flow of water. Nevertheless, they sometimes fail with destructive consequences. Flooding is an odd disaster in the sense that we now seem to be at increasing risk for it, even as we become better equipped to handle almost every other kind of weather.

For severe weather, radar gives forecasters the ability to produce tornado warnings. Satellites show us incoming storm systems on a broad scale. Computer models give us all sorts of time to prepare for significant storms. Plows and road salt make for a quick and efficient rebound after winter storms. Mariners can make use of improved forecasts to keep ships out of harm's way instead of being caught in dangerous waves. In contrast to all these other situations, we may actually be making flooding worse even as technology and knowledge advance.

Why? We go out of our way to disrupt the natural flow of water. Large swaths of the world, which were once forests, are now bustling towns and cities. That might be great for people, but not good for water molecules just looking for somewhere to go. Growth has created a less porous landscape of asphalt and concrete, so when it rains heavily, drains and sewers struggle to keep up with water that otherwise would have soaked into the ground. It is a whole lot easier to flood a city than it is the woods.

Dams help us collect water during a dry period or release it when there is an overabundance. An excellent tool most of the time. However, their formidable power can give us a false sense of security, because what they have really done is give people the impression we are in charge of the rivers behind them. A breaking dam might be one of the single most dangerous events possible for a downstream community. Dams also deny a river from flowing the way it would prefer, disrupting the natural order of things.

People have also added greenhouse gasses to the atmosphere, which, in turn, has been steadily increasing heavy rainfall events. For every 1°F of global temperature rise, the atmosphere can hold 4% more water vapor. The frequency of one-inch, two-inch, and three-inch rainfall events in the US has risen markedly since the mid-1900s, with the northeast region seeing a 55% increase in its heaviest rainfall events (top 1%) since 1958. So when it rains, it does indeed pour. If rainfall comes down at a faster clip, the result is a stress on infrastructure and a higher chance of flooding.

Finally, humans really are just bipedal beavers sometimes. Undeterred, we will rebuild in every imaginable risky place, every time you knock our house down. Rivers provided power for the mills of early industry, but also decided to occasionally splinter and toss them downstream. They make for an inviting spot to build a home or farm but can expand to engulf that scene of Americana. Plus, who doesn't love a seaside home? We know that all these locations carry an element of danger, but we relentlessly plow a field and plant a flag in the face of it. When the local river roars over its banks and takes the house with it, little time is wasted in building it right back. People continually buy homes on the dunes of Cape Cod to watch them fall into the ocean. We are really good at tempting water to take its best shot.

Often, it does. Whether a localized flood after intense heavy rainfall, a tropical storm drenching the region, or a spring freshet. As indispensable as water is, too much of it can cause a world of trouble. We demand it, but in the right proportions at the right time. Unfortunately, nature does not take requests. Here we look at two of the largest floods during modern times, and some of the lessons learned from them.

Connecticut River Floods, 1936

"A great mass of frozen precipitation was stored on the hills and mountains awaiting spring conditions that would be its release to start downhill on the long journey to the sea."
—David Ludlum, The Vermont Weather Book

Before engineers got their hands on the waterways of New England and helped tame their wild flows, spring flooding was a rite of passage. It was assumed that in most years there would be a period of flooding and after the water receded it would be time to move on with the growing season along the road to summer. When looking back it can be said that of all the years before or since, no spring flood has eclipsed what transpired in March of 1936. Far from typical high water, this monstrous flood inundated towns and cities across the whole of New England with a rushing and destructive torrent. Of all historical floods, this is a rare case when all six states participated in the misery at once.

The cycle of the seasons calls for winter's snow to melt away as the sun climbs higher and temperatures slowly warm. In 1936, there was work to be done. It was a notably cold and snowy season with plenty of snowpack to start the month of March. Meteorologists and hydrologists track snow-water equivalent, or SWE, to determine how much water is locked up in the snow. This metric is critical for determining how much can be released into waterways and therefore what the flooding risk is for a given time. It was estimated that in early March, there was about 7.5 inches of SWE across northern New England and 3.5 inches across southern New England. In other words, if all the snow melted at once, the effect would be the same as a sudden 7.5-inch rainstorm dumping down across that area. From the elevated headwaters in northern New England this "trapped" water ready to be released had the capacity to rush downstream with all its might. A combination of milder air and heavy rainfall was about to release that potential.

It all began with the first of several storms on March 9. Spring is well known for slow-moving systems or "cutoff lows" that become displaced from the steering currents of the jet stream. As this guiding upper-level river of air begins to recede to the north, storms can become lost and listless beneath it. The first heavy rainmaker during this stretch was such a storm, lifting a warm front up across central New England and then stalling. Its origins traced back to the Gulf of Mexico, so it was bringing plenty of tropical air along with it. From the 9th through the 13th, consistent rain that could have produced flooding on its own without the help of melting snow pounded down across the White Mountains of New Hampshire.

In what would become the epicenter for heavy rainfall over the course of the month, Pinkham Notch observed 6.46 inches on March 12 and 8.71 inches over the five-day period of March 9 to 13. High rain totals of five inches or more extended eastward into Maine with over two inches in eastern Vermont. Where exactly this initial storm wrung out its moisture played a large role in how widespread the flooding would become. Pouring over the source of many of the region's largest waterways, in an area where there was maximum snow to melt, was the worst possible scenario for a far-reaching disaster.

We saw with the '38 hurricane how forecasts could fail in this early age of weather prediction, but the US Weather Bureau was out in front this time. On March 11 a warning was issued for flooding due to the forecast of heavy rain and the likelihood of significant snowmelt. Getting the word out likely saved lives, and at least gave people a fighting chance to prepare. At the very least, they were on their toes and ready for action.

The warning certainly verified as the Merrimack, Androscoggin, Kennebec, Pemigewasset, Connecticut, and all their tributaries quickly rose. At the start, many were still frozen after a harsh February. As streamflow quickened its pace and built up underneath, the ice started to break up and these chunks of winter debris would set course downstream as destructive projectiles, crashing into infrastructure on their way. Ice jams began to build where kinks in the rivers collected them, giving way when the building pressure of the water behind them became too great. It all played out as a great dance between forces. A swelling flow of water breaking up ice, which then produced jams, only to give way again, pushing an even stronger current downstream.

The jams hit Maine particularly hard, with eighty-one state highway bridges needing to be rebuilt after being punished by the angry waters. When they gave way, record flows resulted. The Androscoggin reached an all-time peak flow in Rumford and Auburn.

A particularly memorable ice jam on the Connecticut River in Holyoke, Massachusetts, made for a spectacular sight. At first, the buildup of water found a way around by cutting a new channel on the east side of the river. When ice above the Holyoke Dam broke and released on the evening of March 15, huge chunks in the rushing water crashed into the granite structure. So devastating was this power that it sheared a piece of the dam a thousand feet long by five feet high right off while water more than nine feet deep overtopped the dam.

Initially this first salvo of ice breakup and heavy rain brought all rivers up into flood, but the hardest hit spots were generally along small tributaries and streams. While damage was significant, it acted more as a primer for the harsher flooding to come. The West River in Brattleboro, Vermont, washed out Route 30 and forced schools to cancel on March 12. Ice chunks that had washed up on the roads blocked travel. Similar scenes played out in New Hampshire and the interior of Maine. During a break in the rainfall on March 14 and 15, these smaller but swollen streams receded somewhat but the larger arteries of the Connecticut and Merrimack continued to rise as rainwater and snowmelt made its way into these main highways of the New England water cycle.

Then came a second strong storm on March 16 to 19. This too was a sluggish, drenching system bringing an even more exceptional rainfall. The footprint was much larger than the first across central and southern New England. Back in Pinkham Notch, a staggering 11.62 inches of additional rain fell. The orographic lift of the White Mountains helped maximize rainfall, though lower elevations also saw more than could be absorbed. For this second event numerous reports of rain totaling two to five inches were noted across much of New England. These locations include the Merrimack and Blackstone Valleys, where all-time flooding was about to commence. A weaker but still soaking storm followed up on March 21 and 22 to add the final blow, prolonging the flooding ahead by not allowing rivers to recede.

At this point, so much water had come together in so short a period that there was no holding it back. A post-storm analysis by the US Geological Survey indicates widespread six- to fourteen-inch rainfall for the full two-week run from March 9 through 22. More than 1,500 square miles of land across the northeast saw at least a foot of rain. The leader of the pack was Pinkham Notch with a whopping 22.43 inches observed. Temperatures were turning unseasonably warm with highs in the fifties and sixties pushing into northern New England. Factoring in snowmelt, which was total outside of high peaks and northern Maine, that brought the tally to thirty inches of liquid over fourteen days. Now came catastrophe.

All previous records were swiftly washed away on numerous waterways. In the southern New Hampshire town of Hooksett, water eighteen to twenty feet deep blasted through Main Street. Barns, schools, homes, power poles, and several hundred feet of railroad embankment twenty feet high were all washed away. The generator station at the Hooksett power plant was demolished.

Just south in the city of Manchester, damage to the Amoskeag Mills was severe. Records kept at the mill since 1785 indicate that the Merrimack had never come within 6 feet of the water level witnessed in March of 1936, which reached 87.1 feet. Throughout the Merrimack Valley there was chaos as more mills, factories, and towns went under. Lawrence, Lowell, Haverhill, and Manchester, New Hampshire, all felt the force of their worst flood on record topping crests from great floods in March of 1896 and November of 1927. Nearby Nashua quickly went underwater as well as the Souhegan River overtopped its banks. Author Jack Kerouac would later describe the flood as "an unforgettable flow of evil and of wrath and of Satan barging through my hometown" in his novel *Dr. Sax*. His father's print shop in Lowell was destroyed in the flood.

To the south in Massachusetts, the Nashua River was raging. The North Nashua set new high-water marks in Fitchburg and turned Lancaster into a temporary pond. On March 19, the *Clinton Daily Item* wrote "It is a weird and awesome sight in Lancaster, today, as the town is virtually an island, the only opportunities for its citizens to reach Clinton being to walk across the old bridge of the Worcester Consolidated Street Railway."

Draining all western New England to the sea, the Connecticut River bulged to levels never seen before or since. All the rain falling across Vermont and western New Hampshire rushing in from raging tributaries was joined by rapid snowmelt. The result was a catastrophic flood that led to the evacuations of numerous towns and left thousands homeless in the aftermath.

Starting first in Vermont and then gaining depth and damage potential during the journey south, the Connecticut River swept away bridges and inundated vast swaths of land on either side of its banks. In Northfield, the Central Vermont Railroad Bridge was lifted, as if effortlessly, from its stone footings and sent downstream. A gorgeous eight-hundred-foot-long bridge in Montague, so well-known for its engineering that the plans were kept in the Library of Congress, was also picked right up by the rushing water. But it did not sink, which made for a memorable sight downstream. Made of wood timbers, it kept floating by to the astonishment of onlookers.

A little stretch farther downstream, the towns of Sunderland and Hadley were both evacuated. There are oxbows in the typically gentler flow of the river in the Pioneer Valley, and they provided excellent breeding ground for more ice dams full of chunks the size of cars. An extremely large one piled up near Holyoke at the base of Mount Tom. An ice dam fifteen feet high was the result, and when it gave way the roar of water breaking free and rushing southward could be heard for miles around. It can only be imagined how terrifying such a sound could have been for those living downstream.

Many of those people lived in the city of Springfield. When the river arrived with all its force, neighborhoods were quickly turned into lakes. On March 18 it broke through a dike protecting West Springfield where eighteen miles of city streets went underwater, and the best mode of transportation became a canoe. Throughout all the towns in western Massachusetts that lined the Connecticut River, fifty thousand people were displaced and left to look for shelters. Doctors made rounds, helping to vaccinate for typhoid that was a risk in the floodwater. From Northampton and Holyoke to Springfield, the river would crest late in the day on March 19 into the morning of March 20 and those high-water marks remain today.

One more state was set to bear the full force of intense flooding, and that was the river's namesake. Connecticut. What initially began as a tourist attraction as many gathered to watch the spectacle became a serious threat to life and property. On March 21, during an all-time crest of 36.7 feet, the capital city of Hartford saw one-fifth of its land and businesses become part of the Connecticut. There wasn't any way to let the rest of the world know, because the entire phone system went down when Southern New England Telephone's main office was flooded with twenty feet of water. In the coming days switchboard operators came back to the building by way of rowboat to restore service.

Electricity was in demand though supply was impossible. The power plant had also gone under and power was out throughout the city. Employees of the *Hartford Courant*, attempting to document the disaster, wrote their stories with the assistance of candles and flashlights. Doctors diligently attempting to go about their work performed operations at Mt. Sinai hospital using battery-powered searchlights to illuminate their patients. Nearly one thousand National Guardsmen were deployed at the height of the flood, patrolling by boat to help with relief efforts and to keep a lookout for vandals and looters. As chaos took over, the city descended into martial law.

The Park River, which used to run through Bushnell Park in downtown Hartford (it was later buried by the Army Corps of Engineers to prevent future floods), became indistinguishable from a large lake in the middle of the city. Water came up to the front of the state capitol building but did not make it inside. A dike protecting the nearby airfield, Brainard, gave out and water quickly submerged hangars to their rooftops. The world all around seemed to be made solely of water and there was no getting in or out by any method other than floating vessel.

To that end, aid in the form of the Coast Guard arrived. Their cutters sneaked into neighborhoods on the high water and brought trapped families to safety. The Red Cross mobilized by setting up in the Old State House to take on storm refugees.

Farther downstream to Middletown, a *Hartford Courant* account would describe the apocalyptic scene of debris threatening to take down Middletown-Portland highway bridge and railroad structure:

"by nightfall both trembled under the constant battering of debris which the racing currents continually hurled against them. Driftwood and lumber, barrels, parts of buildings and even entire houses and barns, furniture, and great oil tanks pounded the bridges with all the terrific force that can be exerted by a mass 300 feet wide, 100 feet across and 10 feet deep."

To save them, crews threw dynamite into the debris that threatened to push the structures down with the force of the river behind it. The debris did indeed break up, and the bridges narrowly survived the flood.

There was also an ecological disaster taking place. Expanding water was collecting all manner of gas, oil, construction materials, bits of homes, and sewage and tossing them throughout the watershed. Eventually, all of it would end up in Long Island Sound.

The catastrophic damage along the banks of the Connecticut hit numerous cities and towns, but it was not the only river in the state running amok. To the west, the Farmington River dealt a devastating blow to the town of New Hartford. On March 19 the entire thirty-two-foot-high by two-hundred-foot-long span of the Greenwoods Pond dam was destroyed. This immediately sent a wave of water eight million gallons strong into the town, wiping out a large part of the downtown and crushing dozens of houses. As soon as the dam gave way, warning sirens blared out. Townspeople took the warning to heart and quickly ran for higher ground. Still, one hundred families lost their homes.

An estimated 150 to 200 lives were lost in the floods, but the loss could have been so much greater. An accurate warning from the Weather Bureau helped get many out of harm's way before the worst of the flooding arrived. Total property damage across New England totaled $100,000,000.

Like so many major weather events of the past, the 1936 flood brought about change. During the Depression, everyone was looking for a job and President Franklin D. Roosevelt used the disaster to get them to work. Several months later, on June 22, he signed the Flood Control Act of 1936. A gargantuan undertaking, it gave the green light for the Army Corps of Engineers to set about preventing future floods. This plan included hundreds of miles of levees and flood walls to be constructed, as well as the creation of reservoirs. In fact, 375 new major reservoirs were to be dammed up. It would take a while for the work to be finished, because amazingly a flood nearly as terrible followed just two years later. The Hurricane of '38 brought the Connecticut up to within two to three feet of the crests witnessed in '36. But once this terrible time passed, work was indeed completed and saved billions of dollars in future flood losses. You can see how well the plan worked today, because the two greatest floods ever witnessed on the Connecticut remain '36 and '38. Nothing has come close since.

The 1955 Floods of Connie and Diane

"I was standing at the corner of Main and Maple streets, by the bridge, watching the police set up barricades, when the bridge collapsed—with a car on it—into the river."

—Ron Greski, Ansonia resident

There are many complex and fascinating elements out there on good old planet earth. If you had to pick the most remarkable one though, it just may be the most basic-looking and common molecule around. Water. It is a remarkable thing. Water exists in all three states on earth (liquid, gas, solid). You can dissolve more substances in it than any other material we know of. It expands when it freezes, allowing it to float atop the liquid form of itself. Life as we know it would not exist without water. And would you believe all the amazing atmospheric phenomenon we see on a daily basis is produced by just a tiny fraction of the earth's supply? About 0.001% of all water on the planet is in the air around us, as a vapor.

On the other hand, water can be a menace. Anyone who has owned a home can tell you that. Just a little leak can wreck part of a home and become extremely costly. We go to great lengths to keep water out of our living spaces even as we depend on it to survive. When large quantities of it conspire to get together and pour down from the heavens, there is nothing quite as destructive as water.

This brings us to the summer of 1955. New England had just gone through one of its worst hurricane seasons the previous summer with Hurricane Carol and Hurricane Edna. The year before that, the Worcester tornado. Unfortunately, the region was having a terrible streak of weather luck, and it wasn't over. While Carol and Edna were direct hits that brought major winds and powerful storm surge, the 1955 duo of

Connie and Diane delivered a deluge. It would turn out to be one of the worst floods New Englanders have endured dating back to colonial times, and earned the unwelcome title of the first billion dollar hurricane disaster on US soil.

This performance was a duet, and stepping onto the stage first was Connie. A tropical disturbance developed in early August and spun up into an impressive Category 4 hurricane with 145 miles per hour winds as it passed north of Puerto Rico. Fortunately for North Carolina, the storm diminished considerably before reaching the coastline and came ashore on August 12 as a far weaker Category 1 with peak sustained winds around eighty miles per hour. From there it slowly chugged up the Chesapeake and then curved to the northwest across Pennsylvania toward the Great Lakes, losing tropical characteristics as it pushed farther inland.

Connie produced major flooding, but this initial round focused most intensely on the Mid-Atlantic to New York. Over a foot of rain fell in Fort Schuyler, New York, and amounts over ten inches were common across eastern North Carolina northward along the Chesapeake and into eastern Pennsylvania. Flooding was particularly intense along the Pamlico River in North Carolina, where National Guardsmen had to evacuate 1,000 people during the storm.

In New England, the heaviest rainfall set up across western Connecticut and Massachusetts. An impressive five to ten inches of rain were observed, including 9.02 inches in Norfolk, Connecticut. To the east, a general three to six inches of rain came down across the Connecticut River Valley. It was not enough for widespread significant flooding but was successful in bringing local streams and rivers above flood stage. In the tale of the '55 floods, this first rainfall was a critical primer. The ground quickly became saturated after what had previously been a hot and dry summer. Limited ability to take on additional water due to antecedent conditions is often a staple of some of the worst floods, and this situation was no different.

The 1950s still did not have the critical tool of satellite imagery to fully assess what was going on out over the Atlantic, but if it had been available the sandbagging operations would have been full-throttle in the wake of Connie. That is because another hurricane was brewing, and it was not far behind.

A wave off the coast of Africa gradually organized into a tropical depression on August 7, and two days later Tropical Storm Diane was born. After another two days Diane would reach hurricane status and seemed intent on following in Connie's footsteps. It peaked as a Category 2 hurricane north of Puerto Rico, then set its sights on the North Carolina coast. Fortunately, cruising through the wake of another tropical system isn't conducive to strengthening as the predecessor leaves a trail of cooler water behind it, due to upwelling. Diane came ashore as a strong tropical storm just five days after Connie; and just a mere 150 miles farther south near Wilmington.

Diane was not a classic powerhouse storm, and the full scope of what it could do was not yet clear for many forecasters. In Connecticut, it was not widely thought that the biggest flood in the history of the state was about to commence. The Weather Bureau would later admit that they "goofed" and publicly admitted they made a huge mistake. Once Diane had moved inland they did not anticipate it curving back over open water south of Long Island and packing a devastating punch of rain. The forecast called for "scattered thundershowers."

Instead, what followed was an astounding amount of rainfall. Between August 18 and 19, the ragged center of the storm emerged from the coast of New Jersey and paralleled the southern New England coast. Over a twenty-four-hour period, records were shattered for both rainfall and discharge on numerous rivers and streams. Water came up to levels not seen in over one hundred years. The effect for many towns, homes, and families was immediately devastating.

Where Connie's axis of heaviest rain took on a north-south orientation maxing out along western New England's high terrain, Diane's axis was oriented west to east along a zone of strong convergence north of the cyclone's track. Tropical moisture streaming up from south of Long Island was being lifted over this front and wrung out like a sponge. The overlapping region between the two storms was a section of northwestern Connecticut into western Massachusetts, and the rain totals over an eight-day period were more than anyone could have expected.

Rain Totals from August 12 to 20, 1955

Westfield, Massachusetts 25.87 inches

Barkhamsted, Connecticut 25.06 inches

Burlington, Connecticut 24.65 inches

Norfolk, Connecticut 21.81 inches

Of the total in Westfield, a whopping 19.75 inches fell during Diane's downpours, nearly half of what is typically seen in an entire year. The Westfield River exceeded the previous high-water mark by five feet. In Connecticut, swollen waters quickly set new record marks on the Farmington, Quinebaug, Housatonic, and Naugatuck Rivers.

A passage from the preliminary Weather Bureau report described a record-setting rainfall they observed at Bradley Airport in Windsor Locks, Connecticut:

"Until about 9 p.m. on the 18th, the intensities fluctuated considerably, but from then to 10 a.m. on the 19th the rate was quite constant, averaging nearly .6 inch per hour for 15 hours. The greatest amount from this record in a 24-hour period, 12.05 inches, is from 10 a.m. August 18 to 9 a.m. August 19. This compared with the previous maximum 24-hour rainfall record at Hartford, Conn., of 6.82 inches occurring on July 13, 1897."

It still holds the twenty-four-hour record for that site, as of this writing. The total for the month at Bradley Airport was 21.87 inches of rain, which also stands (and more than 5 inches above second place, October 2005).

With the sopping-wet soil and river levels high from Connie's rainfall, there was nowhere for the excess water to go. Where terrain came into play, the added velocity of rivers rushing downstream roared through towns and over streets. The sheer power of it blasting wood and stone out of its way.

Across western Connecticut, it was catastrophic. Flooding on the Naugatuck River wiped out so many bridges that the state was effectively cut in half for a time with little to no communication traveling across it. Torrington, where the main stem of the river begins, saw 16.86 inches of rain in twenty-four hours. Maximum rainfall coming down at the head-waters of a large river is just about a worst-case scenario. For approximately forty miles the whitewater destroyed property on its journey to join the also raging Housatonic River in Derby.

The town of Winsted saw its commercial district crushed by the raging waters of the Mad River, which came right down Main Street. In an instant, 95% of all the businesses were wiped out. The damage was so terrific that Winsted never fully rebuilt and eventually paved over the street where buildings had once stood. In an article for the *New Yorker*, author John Hersey described the apocalyptic rush of water.

> *"Along Main Street, it was 15 feet above its normal level, and the water was 10 feet deep in the street itself. It was literally ripping up Main Street. The pavement and the sidewalks were being sliced away and gutted six feet deep. The water had broken the plate glass windows of most of the stores along the street and had ruined their stocks. Winsted Motors, a Buick showroom and service station that had straddled the river high up the street, had been completely demolished and its new and used cars were rolling all the way downtown, and its roof had lodged itself in midstreet right in front of the town hall."*

The National Guard was activated, and helicopters were deployed to rescue people off the tops of rooftops and trees. During a flyover Lt. Col. Robert Schwolsky would describe the surreal scene from above: "I've never seen anything like Winsted's Main Street. It looks like someone had taken cars and thrown them at one another." During the flyover they also spotted a house, floating downriver, with the chimney still smoking.

Downstream in Waterbury, twenty-six people were killed when the torrent swept away thirteen houses on North Riverside Street. Other towns succumbing to unprecedented damage included Collinsville, Unionville, Ansonia, Thomaston, Farmington, Suffield, and New Hartford.

But it was not all in western Connecticut. To the east in Putnam, the Quinebaug River produced widespread damage that also led to a huge fire at the local magnesium-processing facility. Hundreds of barrels floating away down the road exploded, creating a spectacular sight.

The Governor of Connecticut, Abraham Ribicoff, declared the floods "the worst disaster in the state's history." President Eisenhower agreed, and on August 20 declared Connecticut a "major disaster area." Seven other states would join that list in the coming days. The toll had certainly been great. Of the 180 total lives lost during the floods, seventy-seven of them were residents of Connecticut. It is estimated that nearly seven thousand were injured across New England. The price tag for infrastructure loss came in at over $350 million, much of it commercial properties and bridges. A state report several months later would note 2,460 homes suffered major damage while 668 were destroyed. More than two hundred dams, mainly located south of Worcester, Massachusetts, failed during the widespread flooding in the wake of Diane, which in turn cost one hundred thousand people their jobs when the mills and factories dotting the shores of rivers were suddenly gone.

Connecticut featured the greatest loss, but damage extended across state lines into Massachusetts and Rhode Island as well. Though a difference in terrain changed the type of destruction brought upon these towns and cities. The Blackstone Valley is a little more rolling hills than the steeper inclines found in western Connecticut, and so inundation took center stage over the velocity-driven power that blew apart towns to the west.

The Blackstone River swelled to a mile wide and in the Rhode Island city of Woonsocket doubled the previous record for streamflow that had been set during the Hurricane of '38. An earthen dam by name of the "Horseshoe Dam" would saturate under 10.43 inches of rain and give way, devastating areas downstream. A rushing surge of water twenty feet high swept out of the Harris Pond that it had been holding back for ninety years. A grim outcome followed as the water picked up coffins from cemeteries and carried them downstream.

Along the French River that runs from Leicester, Massachusetts, southward to the Quinebaug in Connecticut, nearly all dams were obliterated. Widespread flooding swept through the city of Worcester, with record flooding noted on Kettle Brook. More records were set on the Assabet River in Maynard and the Taunton River as well. Part of a passenger train on the Boston and Albany railroad plunged into the Westfield River when the roadbed it was traveling along caved in.

In an iconic spot from the American Revolution, the Old North Bridge in Concord was demolished. A replica would replace it the following year.

Few living or working near a running body of water were spared by the flooding, and repairs would be slow to come. Unbelievably, another flooding rain event was just a couple of months away. During October, a four-day storm would bring another ten to fourteen inches of rainfall and some rivers, such as the Housatonic, would reach or surpass the records just set during Diane. There has never been a year in southern New England with worse flooding than what was witnessed in 1955.

The legacy of the Worcester tornado was the reorganization of the Storm Prediction Center and deployment of more radar to detect storms. Connie and Diane left a legacy as well, and that legacy is flood control. In the coming years, the Army Corps of Engineers went to work and tens of millions of dollars were spent to construct more dams and flood control measures throughout New England.

The Colebrook River Dam would became a major part of flood control on the Farmington River after its construction by the Corps. Winsted received more protective measures with the Sucker Brook Dam and the Mad River Dam to hold back the unruly flow of the Mad River.

Seven new dams were installed along the Naugatuck, which had flooded every single town along its banks. The French and Quinebaug Rivers both received more dams and connections to nearby ponds to control the flow of water during high precipitation events. And to the south, the Thames River Basin received six new dams to significantly decrease the threat of major flooding.

To be sure, any storm system prolific enough to drop two feet of rain over the course of a week will still cause widespread issues across the region. But the addition of numerous dams and flood-control measures, as well as an expanded culvert and drainage system, greatly improve the ability to manage and respond to these high-end rainfall events in the wake of '55.

The national flood program took a long time to officially get started (1968) but its slow beginnings were started after Diane. In 1956 Congress passed the Federal Flood Insurance Act, which created a federal flood insurance program. They left out the important aspect of actually funding it though, so it would be over a decade before the initiative became reality. After so much destruction and financial loss during the back-to-back hurricane seasons of 1954 and 1955, public outcry also led to the formation of the modern-day National Hurricane Center in Miami, Florida, during the summer of 1956.

As a postscript, it may be important to note that neither Connie nor Diane made an actual landfall in New England. There is a lesson to be learned here when preparing for future storms. It is easy to get caught up in the peak winds of a tropical system and the path of its center. The focus is on category, and everyone looks at these systems as wind machines capable of ripping off rooftops and throwing debris. Certainly, this is a hazard. Whipping winds affect a few, but torrential rainfall fans out over a much larger area. Though there are a myriad of ways tropical cyclones can impact a region, and without a doubt the immense quantities of water they carry is a big risk as we have seen repeatedly over time, including August of '55.

Tropical Storm Irene Flooding and 1927 Vermont Flood

"I got my camera to take pictures, knowing people would not believe the devastation without seeing it for themselves."
—SERGIO PEDRINI, ROXBURY, VERMONT

TROPICAL STORM IRENE FLOODING

IT IS HARD TO IMAGINE ONE OF THE REGION'S MOST DESTRUCTIVE weather events of all time being discussed as a "bust," but that is exactly what was going on at the end of August in 2011. The possibility of a massive storm surge flooding out New York City's subways and tunnels came up a few inches short, and with millions of eyes on the biggest city in the country it felt as if Tropical Storm Irene had failed to live up to expectations. Much to the contrary, while this discussion was taking place there was an epic disaster of generational proportions tearing apart Upstate New York and western New England. Irene is the ultimate example of how there is no such thing as "only a tropical storm."

Winds were feisty enough for widespread tree destruction and long-lived power outages. But it was water that made headlines in the end. Swelling streams gained momentum and swept away homes, lives, and historic covered bridges. Water in very unfamiliar places surrounded towns and cut them off from the outside world. The flooding would destroy 3,500 homes and businesses and put 20,000 acres of farmland underwater in Vermont alone, where 225 of the state's 251 towns reported damage to public infrastructure.

Even with New York City avoiding a worst-case scenario (it would arrive the following year with Hurricane Sandy), the Connecticut shoreline was inundated with some of the highest water since the '38 hurricane. A combination of the three- to six-foot storm surge arriving at high tide, a new moon producing extra-high water levels, and a contribution of sea level rise between 1938 and 2011 brought extreme flooding into Long Island Sound. Dozens of homes were wrecked on the Connecticut shoreline with the hardest hit communities found on the western end of the Sound. At this bottleneck, water piled up and was subsequently shoved inland on southerly winds as the storm passed. Branford, East Haven, Milford, Guilford, and Bridgeport took the brunt of the surge. There was an extensive trail of destruction left behind and it left many scratching their heads wondering how a tropical storm could be so punishing.

The system began as a classic African easterly wave that swirled its way south of the Cape Verde Islands on August 15, kicking off the long sojourn across the Atlantic. Slowly organizing, it earned the name Irene about 120 nautical miles east of Martinque on August 20. Irene reached hurricane status while crossing Puerto Rico and, once clear of Hispaniola on the 24th, exploded into a dangerous Category 3 hurricane in the Bahamas. Anything tropical in the Bahamas is worrisome for New Englanders, and that old adage would unfortunately prove true yet again.

Meteorologists at the National Hurricane Center, who had been tracking the storm since its leaving Africa, issued a near-perfect forecast early on the morning of August 25. It called for Irene to make landfall on the Outer Banks of North Carolina and then move northward to New York City while weakening to a tropical storm and continuing across western Connecticut and Massachusetts. This is precisely what played out. In all, there were several landfalls on the eastern seaboard including Cape Lookout, North Carolina; Brigantine Island, New Jersey; Coney Island, Brooklyn; and then finally over Manhattan, New York.

On the eve of the storm's arrival in the northeast, it felt like the world came to a standstill. An unprecedented stoppage of transportation systems put a temporary halt on travel whether by train, plane, or automobile. New York City shut down its subway system for the first time ever due to a natural disaster. All three New York City–area airports closed on August 27 and did not reopen for two days. The masses hunkered down and waited for the tempest to unfold.

It was a dark and stormy Sunday afternoon and night, August 28 into the 29, when Irene sliced across New England. Flood watches escalated to flood warnings. A widespread two to five inches of rain pelted down and drenched the landscape. Across the higher terrain, orographic enhancement squeezed out even more. A dangerously intense rainfall of six to eleven inches swamped southeast-facing slopes, creating a wall of water racing through the Green Mountains, Berkshires, and the Catskills of New York. In a vacuum without any other compounding variables, flooding would be a significant problem. But the storm was about to elevate into the upper echelon of New England floods thanks to what came before it.

Ingredients to make Irene a devastating force had been gathering weeks ahead of landfall. Rain was falling in enthusiastic fashion all month and saturating the region. From the Mid-Atlantic to New England, soil moisture ranked in the ninetieth percentile compared to long-term averages. The most drenching antecedent conditions were in New Jersey, where eight to sixteen inches of rain had already come down during the month of August. To the north totals had been less extreme, but complex terrain would make up the difference. Poughkeepsie, New York, observed just over six inches for the month leading up to Irene with a little over five inches in Rutland, Vermont. These amounts are still more than average for all of August, and wet soil was a problematic table-setter. When the fast-falling tropical downpours of Irene arrived, the ground was incapable of absorbing significant rain. The result was a catastrophic amount of runoff.

No place would garner more attention or see more devastation than the state of Vermont. In a matter of hours water exceeded or rivaled records set in the historic Great Flood of 1927. Everyone expected heavy rain and flooding, but no one truly thought they would wake up to a world turned upside down. The surreal transformation sent homes floating downstream on a number of swollen rivers. Boulders and debris tumbling in the torrents scoured out riverbeds and demolished anything in their way. Roads were washed out and closed in seemingly all directions. Once familiar waterways were unrecognizable in their new, vast scope.

Overnight, thirteen communities had been completely cut off from the outside world. Floodwaters surrounded them and washed out bridges that previously granted access to other towns. A staggering amount of these bridges, 480 strong between state and municipal jurisdictions, were ripped from their foundations, damaged, or sent floating in the torrents. The majority were the standard highway bridges you might see anywhere in the northeast. It was the loss of quintessential Vermont covered bridges, however, that broke hearts.

The 1927 flood took out nearly half of the state's most vulnerable covered bridges, which numbered two hundred at the time. Irene did not match that terrific loss, but four of the final one hundred were taken or severely damaged by the storm. In Rockingham the beloved Bartonsville Bridge was battered by the Williams River. Ironically, the bridge had initially been built to replace one destroyed in the October flood of 1869. For 140 years it had survived the floods of 1927 and 1938 and earned a place on the National Register of Historic Places. Irene delivered the fatal blow. After abutments were scoured out and gave way, the 158-foot bridge set sail down the Williams and eventually flipped on its side about a half mile downstream. Video of its demise, captured by onlookers, would become one of the most widely-played scenes from the storm's rampage.

Roughly thirty-five miles due north, another iconic bridge took a severe beating from the Ottauquechee River. The picturesque Quechee Bridge is a wonderful welcome mat for Quechee Village, and on a normal day sits comfortably above the Ottauquechee River with the famous Simon Pearce glassblowing operation on its flank. It is a photo millions of travelers have stopped to take. Rainfall from Irene took the river to shockingly new heights and brought out "storm tourists" to gape at the spectacle. Quickly, water rose to the base of the bridge and battered it into submission. The road leading into the bridge collapsed, though the structure did not completely wash away like the Bartonsville. Teetering on its abutments, enough had been claimed by the Ottauquechee that a full rebuild was required in the wake of Irene with some modernized enhancements and an increase in length from seventy to eighty-five feet. Simon Pearce, along with several other nearby businesses, sustained flooding damage but remained on dry land when the river subsided.

The two other badly damaged covered bridges included the candy apple-red and postcard-perfect Taftsville covered bridge, a jaunt down Route 4 in Woodstock, and the Upper Cox Brook Bridge in Northfield

Falls. The former spans an impressive 189 feet and has survived rainstorms and floods since its completion in 1836. It barely made it through Irene and took significant lumps from floating debris. Deemed unsafe, restoration would take two years to complete. The covered bridge above Cox Brook weathered the storm in terms of flooding but was wrecked when a tree was pushed through the bridge and into the roof.

While out of state, another notable loss in the family of historic covered bridges was the Blenheim in Upstate New York. Located on Schoharie Creek, the Blenheim had a reputation of the longest covered bridge in the world at 210 feet. Riding out storms since 1855, Irene was too much and ripped it apart, lost to future travelers forever.

Back in Vermont, the nerve center of disaster response in Waterbury was in trouble. A sudden rise of the Winooski River was conspiring to derail a number of important storm-related operations. The state office complex home to the Vermont Agency of Emergency Management and approximately 1,500 employees was inundated when rushing water and mud engulfed the village. Evacuation orders went out Sunday evening and many state and emergency offices needed to relocate for their response efforts. The Vermont State Hospital, which housed and treated some of the state's most severely mentally ill, was among the many evacuated under chaotic conditions. The resulting damage throughout Waterbury was catastrophic, flooding out two hundred homes and businesses.

Harrowing stories seemed endless with rivers running amok and scenery becoming unrecognizable to many Vermonters. To the south of Waterbury it was the Mad River blowing apart roads and homes in Moretown, Waitsfield, and Warren. Continuing down the eastern flank of the Green Mountains, Granville, Hancock, and Rochester were crushed by the rising White River. Another waterway in Rochester, Nason Brook, produced a grim circumstance when it undermined sections of the Woodlawn Cemetery. At least fifty gravesites were disrupted and a disturbing search for remains of the dead was required. Volunteers on foot and dog teams set out and found the bodies of twenty-four individuals and bones of several others downstream.

On and on the path of destruction went with speed being the defining characteristic of how Irene matured from a soaking rain without much wind to the worst flood in nearly a century. The Deerfield River, normally a bucolic stream running parallel to Route 100, may as well have been the Connecticut River. Bursting its banks and swelling across the valley, it jumped eight feet in fifteen minutes when it burst into downtown Wilmington. The Deerfield would damage forty-eight businesses in the quaint downtown, one of them being Dot's restaurant, which anchored a building that had stood by the water since 1832.

This does not represent an exhaustive list of all the losses across Vermont, but a sample to visualize part of the dramatic transformation that took place. At least nine stream gages notched peak flows that were at least one-hundred-year floods or had a 1% or less chance of occurring in a given year. Sections of the worst-hit rivers fell into the five-hundred-year flood category or had a 0.2% chance of occurring in a given year. Otter Creek in Rutland and the Wallomsac River in North Bennington both notched their highest water levels on record. Over five hundred miles of road and two hundred miles of rail were damaged in addition to the widespread inundation of homes and businesses, which numbered over 3,500.

Most national media attention would focus on the dire situation in Vermont, and deservedly so. Though to the south the impacts were similarly severe. In Massachusetts, the observed rainfall was reported as high as 9.92 inches in Conway and 9.75 inches in Ashfield, both towns located in the foothills. The Berkshires and Connecticut River Valley experienced a sudden spike on streams that rivaled Vermont and overtook towns. Most of the rise occurred over the span of just a few hours, leaving little time to react and get out of the way. The Deerfield River in West Deerfield jumped from five feet to twenty-four feet in under four hours. Eight of the sixteen USGS stream gages in western Massachusetts with long-term records set new marks for peak streamflow, including sections of the North River, South River, Green River, Mill River, Housatonic River, and Hoosic River. Wherever a record was not quite set, the reigning first place event was typically the '38 Hurricane or '55 flood. It may come as no surprise that the two wettest Augusts on record in the state of Massachusetts are 1955 and 2011, when two of the most historic floods took place.

Hilltowns in Connecticut saw a similar rapid rise on rivers and streams, though in addition to the flash flooding were severe impacts by way of storm surge and strong winds. After touring the damage the day after the storm, Governor Dannel Malloy declared "it is clear that this has been a major occurrence in all of our 169 municipalities," which either had structural damage, flooding losses, no power, or a combination of the three. Shelters and hotels filled up as storm refugees tried to locate places to stay during the cleanup.

Roaring tropical winds were muted compared to those experienced in more powerful storms like '38 or Carol. However, it doesn't take much to cause problems in a heavily forested and populated area. Gusts of forty to sixty-five miles per hour buffeted Connecticut and those alone produced the biggest power outage in state history, topping Hurricane Gloria in 1985. Without question the heavy rainfall and soaking soil allowed more trees than usual to give up their ground and tip into lines and homes. As they swayed and snapped, over 767,000 customers lost power. The widespread nature of the damage not just in Connecticut but up and down the eastern seaboard meant this was not going to be a one-night candlelit affair. The wait to flip switches and streetlights back on was destined to test even the most patient people. Restoration took over a week in some towns, much to the consternation of residents and businesses. The start of school had to be postponed for a week as a number of districts needed the buildings for storm shelters and a large number of families had no electricity to go about the daily routine. It would end up being an unprecedented rough year for public utilities and their customers as a freak October snowstorm delivered another massive and protracted outage just two months later.

Even with ample warning, the coastal damage was shocking for many homeowners. Once Irene was downgraded to a tropical storm while making its New York approach, the general feeling was that it wasn't going to be so bad after all. Instead, storm surge swamping Long Island Sound blasted apart twenty-five homes in the Cosey Beach neighborhood of East Haven and several more homes were lost on the beaches of Fairfield. Caught in the rush of water, the National Guard had to be called upon for water rescues and evacuations.

While 2011 wasn't done with weather disasters yet, Irene was one of the worst events during a very expensive and ruinous year. Its initial landfall as a hurricane in North Carolina gave it an undesirable ranking as the costliest Category 1 hurricane in US history. The final price tag came in at $15.8 billion. At the time, it was also the sixth-costliest hurricane on record in the US with the majority of the losses due to inland flooding. A total of forty-eight people were killed by the storm, forty of them in the US.

VERMONT'S GREAT FLOOD OF 1927

Memories run deep in New England, and some nonagenarian Vermonters may recall both of the state's worst floods. In November of 1927, a tremendous rush of water equaled what was witnessed during Irene and in many places exceeded it. An astonishing 1,285 bridges were destroyed in the onslaught and eighty-four people lost their lives, including Lieutenant Governor S. Hollister Jackson. For many it is still considered Vermont's greatest natural disaster.

It may have been much later in the year, but the setup taking shape was not very different from August of 2011. Once again, antecedent conditions would play a major role. And once again, the coup de grace would be delivered by the tropics. A reminder that the Atlantic Hurricane season runs through the end of November.

To set the stage, the gales of autumn soaked Vermont throughout October. Rainfall was running 150% to as much as 300% above normal for the month. And so like August of 2011, the ground was fully saturated. Lakes, ponds, and vernal pools were already filled up. To make matters worse, it was long past a killing freeze, which, in many spots, arrived during the first half of the month. With summer vegetation on the way out and leaves off the trees, it would be easier for heavy rain to run off rapidly instead of being absorbed into the ground. Any torrential downpours would head straight into streams and rivers and add to their volume.

The tropical contribution came from the seventh storm of the season, developing near Jamaica on October 30. Heading across Cuba, it would continue on its northerly trajectory until reaching the Outer Banks of North Carolina as a tropical storm on the morning of November 3. A cold front moving in from the west was already helping to funnel its rich moisture source into New England, with rain commencing in Vermont on the evening of November 2 and then picking up on the 3rd as the storm drew closer. People on the ground couldn't know it at the time, but the rain would not stop falling for over thirty-six hours.

Being late in the season, the system was unable to hold on to its tropical characteristics past North Carolina. Tracking into New England it would become extratropical, or something more reminiscent of a weak nor'easter. Destructive winds and storm surge would not be significant variables for the coast. Something that did remain, however, was the moisture-laden air sourced from the Caribbean. And that was about to be wrung out over Vermont.

One other critical element that must be noted is the presence of an unusually potent and cold area of high pressure anchored over the North Atlantic with a secondary center north of New England. Not only did this slow down the storm and prolong the duration, but the cold surface air helped to enhance a region of exceptional rainfall across western New England. Acting as a wedge to help lift the warm and moist tropical air, it was the perfect situation to maximize rain.

The deluge arrived on the afternoon of November 3 when southeast winds were lifted over complex terrain and cold air. In just a matter of hours from the late afternoon into the evening, over seven inches of rain fell in some towns. Northfield reported 7.61 inches in twenty-four hours, an amount generally considered obscene for such a northern locale. Compared to Irene, the footprint of water was more widespread with the majority of towns and villages observing four to nine inches. The US Geological Survey estimated that more than half the state (53%) received over six inches of rain and 1,660 square miles over eight inches before winding down on November 4. The highest recorded total from the storm was 9.68 inches in Somerset, Vermont, with Rutland (8.47 inches) and Searsburg Mountain (8.30 inches) also at the top of the list.

Significant rainfall of three to six inches also fell across western Massachusetts, Connecticut, and New Hampshire where more flooding, though less extreme, took place. The Connecticut River at Springfield, Massachusetts, would notch its highest crest on record, only to be surpassed in the 1936 flood and the '38 hurricane.

On rivers and streams, the effect was swift and catastrophic. Mass flooding took over roads and towns throughout the state, with an epicenter in the Winooski Valley. The town of Bolton, sitting along the Winooski between Burlington and Waterbury, suffered greatly in the flood.

The Bolton Valley Dam broke and swept away a boardinghouse, killing fifteen. Seven others would be killed in town before the water subsided, a total of 55 throughout the Winooski Basin. No river would end up being deadlier than the Winooski during the Great Flood of 1927.

The capital city of Montpelier was overtaken by mud and water. The entire business district downtown was under eight to ten feet of water. A traveler can still walk through town and take in the unsettling sight of a high-water marker installed by the USGS. It simply reads "1927" and stands more than eleven feet above sidewalk level.

Fortunately, the federal government commissioned a flyover to survey the damage, giving us an accurate assessment of what transpired. Ninety aerial photos were taken over the Winooski, White, Black, and Lamoille Rivers and the wreckage was expansive. Bridges, roads, homes, and retaining walls were washed away throughout the area.

Agricultural losses were immense in the wake of the flood. Nearly seven hundred farms were bulldozed by rushing water. A total of 1,704 cows and 7,215 chickens drowned, with 16,404 bushels of grain destroyed. The losses were estimated to fall in the $30 to $35 million dollar range, a small fortune in the 1920s. Adjusted for inflation, the 1927 flood in Vermont alone would surpass a half billion dollars today.

President Calvin Coolidge, a native son of Vermont who was born in Plymouth, would visit the state during the following year. The president's prevailing sentiment is one that still lingers today and showed in the aftermath of Irene's flooding, when community spirit rose to the challenge. He remarked "I love Vermont because of her hills and valleys, her scenery and invigorating climate, but most of all because of her indomitable people."

A measure of flood control would follow during recovery. The US Army Corps of Engineers set about trying to tame the wild flows of the state, building three new flood retention reservoirs in the Winooski River basin at East Barre, Wrightsville, and Waterbury. Another reservoir, Union Village, would be completed years later in 1949 on the Ompompanoosuc River. Considering how close Irene's flooding came to 1927 in numerous locations, it is a lesson that human ingenuity still has its limits in the face of the power of nature.

PART FIVE
Tornadoes: An Infrequent but Destructive Menace

Worcester Tornado, 1953

A tinge of green in the sky, the roar of a freight train. We've all heard the anecdotal signs of an incoming twister. No place on earth sees more tornadoes than the US with over one thousand each year. Every single state has seen a dark funnel snaking its way to the ground. They spin up across flat land, through cities, on mountainous terrain, and over water (waterspouts). It is as if someone created our geography with the express purpose of getting them to form.

Across the Plains and Southeast, they are a way of life and one of the most troubling weather hazards. Rare is an outbreak in New England, but you can usually count on a few with an average of seven per year across the six states. There are enough of them that most schoolchildren can still tell you the safest place to be is in your basement and that you are most likely to see one in the spring or summer.

A tornado is a violently rotating column of air that can handily lay waste to anything in its path. Seemingly unbreakable steel beams can be twisted like straws. Cars thrown through the sky and entire city blocks wiped clean in a matter of seconds. Many natural disasters are violent, but none quite as proficient in quickly destroying human property. There is also the least amount of warning. When a hurricane, snowstorm, or flood is brewing there typically are several days' worth of advance notice to prepare. Forecasters may be able to give everyone a heads-up that severe weather is possible days in advance, but truly knowing if a tornado is coming to your town can vary from more than twenty minutes of warning to no warning at all in the worst-case scenarios. Putting all tornado warnings together, the average lead time is just thirteen minutes.

The mid-latitudes of the world are where tornadogenesis is most likely to take place, across a band between 30 and 50°F north or south. It is a "sweet spot" where polar air can clash with moist subtropical air and wind shear is common. These are the key ingredients to watch for when trying to sniff out twister risk. Instability allows air to rise rapidly, cooling as it is lifted and condensing into rain, hail, and towering thunderstorms. You can generate it by warming up the surface, cooling the middle layers of the atmosphere, or both. Wind shear refers to either veering winds with height or increasing speed with height. If, say, winds are south-southeasterly at the surface and westerly several thousand feet up, you will generate directional wind shear. Or if the surface winds are ten miles per hour and increase to sixty miles per hour with height, then you have speed shear. Overlap this shear with instability and you have good odds to create a rotating updraft into a storm.

Figuring out just how many tornadoes there have been across New England is a tricky exercise. For one, they are usually small and brief. Someone must see or be affected by one for meteorologists to know for sure if a tornado occurred. There is a lot of forest across New England and, especially before very recent times, many went undocumented. The historical record is also full of "tornadoes" that were in fact likely microbursts, a straight-line blast of wind that can be equally damaging. After pioneering work by Dr. Ted Fujita from the 1950s to the 1970s, storm surveys became more common and now the cataloging of these events is much more accurate. Our most complete database only dates back to 1950.

Dr. Fujita also gave the world the Fujita scale for estimating the strength of tornadoes based on the damage they leave behind. Updated in 2007 to the "Enhanced Fujita" scale, it ranges from EF0 (sixty-five- to eighty-five-mile-per-hour gusts) to EF5 (more than two-hundred-mile-per-hour gusts). The vast majority of tornadoes in New England are EF0 or EF1 twisters, on the bottom end of the scale. The average number of "violent" EF3 or higher tornadoes per year is a fractional shade above 0, with a total of twenty over the past seventy years (1950 to 2020). An EF5 has never been documented in New England, though at least one came extremely close.

Even in a place where they are uncommon, tornadoes can occur with life-threatening and destructive consequences. One of the deadliest tornadoes ever to hit the US was right here in Massachusetts. Thanks to the dedicated work of severe weather specialists, outlooks and warnings have come a long way over the years. Improvements to Doppler radar have dramatically increased our ability to monitor storms and see their inner workings. But when it comes to avoiding them, there is still little that can be done to prevent their intense power from wreaking havoc.

The Worcester Tornado, 1953

"I was sucked right out and flew down the hallway . . . I was able to hang on to a radiator pipe, and that is where I stayed until it was over."

—RICHARD DION, AGE SIXTEEN,
STUDENT AT ASSUMPTION PREPARATORY SCHOOL

WHEN SOMEONE THINKS ABOUT THE WORST WEATHER NEW ENGLAND conjures up, the mind immediately wanders to images of punishing nor'easters, whipping snow, and damaging surf. Tornadoes? You may have heard of one or two but it is a far cry from Kansas. That idea may have held up for a while, until one steamy afternoon in 1953. On June 9, the atmosphere stirred up images from Tornado Alley and dropped them right down onto an unlikely spot—the state of Massachusetts. Even more unlikely? It was the first full year that public tornado warnings were to be issued.

For nearly an hour and a half, one of the worst tornadoes in US history ripped up everything in its forty-six-mile path from Petersham to Southboro. Trees shredded and snapped, cars thrown across town, whole neighborhoods leveled. This unimaginable rampage took 94 lives, injured 1,300, and destroyed a whopping four thousand buildings before lifting back into the sky that delivered it. At the end of the day, 10,000 people had lost their homes and were left wondering how such a thing could happen here, of all places.

What darkened the skies of central Massachusetts that day was not a localized, one-off event. There was a buildup to atmospheric crescendo for days as the same storm system was wreaking similar havoc across the country. The Worcester tornado was to be the finale.

It all began with a relatively disorganized storm system on June 7 when energy ejecting out into the Great Plains spawned the start of a three-day severe weather outbreak. The saddest tale from this initial burst came from a patch of farmland in the small town of Arcadia, Nebraska.

That afternoon, a family just like countless others on the Plains was getting together for good conversation and a meal. Mads Masden and his wife, Minnie, had invited their three children and five grandchildren over for Sunday supper. It was an extra-special meeting because Mads had just gotten out of the hospital and so the kids were coming over to offer some support. Much like any close American family would do.

The five adults and five children did indeed have a storm cellar available to them on the property, but in a time without up-to-the-minute forecasts and warnings they were unaware that a huge twister was rolling across the farmland toward them. Ripping into the house before anyone made it to shelter, the entire family was killed. Out of thirty-two tornadoes on this first day of the outbreak, the only deaths came in Arcadia. A reminder of the indiscriminate nature of tornadoes.

To the east, word was out. A strong storm system capable of destruction was on the way. In Michigan, they took steps accordingly. One piece of technology that still has not gone out of style is a weather balloon, and the Weather Bureau in Mount Clemens (just northeast of Detroit) launched one that afternoon to test the waters of the atmosphere. A weather balloon, or "radiosonde," produces a picture of the air column tens of thousands of feet up above a given location. Temperature, humidity, and wind speed are all recorded before the balloon pops and falls back to earth. We call the resulting image a sounding.

Particularly in the days before computer models and satellite data, this was invaluable. Even though meteorology was still a budding science, forecasters at the Weather Bureau office knew that tough times were ahead when they saw the information gathered by this launch. All the ingredients were there—unstable air that would easily and rapidly rise, deep wind shear to create rotating storms. They issued a "Severe Weather Bulletin" that looks similar to today's Tornado and Severe Thunderstorm Watch boxes around 7:30 p.m., but by this time tornadoes were already on the ground.

One of these stands above the rest. With 116 killed, there hasn't been a deadlier tornado in Michigan since the F5 that struck the community of Beecher. In fact, it was the only tornado in the US to kill more than one hundred people until 2011, when an EF5 tore apart the community of Joplin, Missouri.

Needless to say, Boston-area forecasters had a lot to digest when they woke up on the morning of June 9 and saw the front pages. They had seen the sensational reports from Michigan. It was hot, it was humid, and they knew by looking at the information at hand that it was going to be a busy day.

It may be difficult to picture now, but in the early 1950s it was almost taboo to mention severe weather. In fact, the word "tornado" had never been mentioned before in a forecast for the northeast. Even when it looked like there might be a chance.

To go even further, the Boston office of the Weather Bureau had yet to issue any sort of advisory or warning including the phrase "severe storms."

At the time, early forecasters were hesitant to scare the public. Anything that might cause panic was frowned upon. And so the morning meeting inside the Weather Bureau office at Logan Airport featured a lot of hand-wringing and lively conversation. It is not as if big tornadoes were frequent, or even remotely common, in the state of Massachusetts. These were monsters that tormented the Great Plains, Midwest, and Southeast. Surely the devastating storms tearing up homes and lives in places like Nebraska, Kansas, and Michigan would weaken before visiting New England?

Enter Al Flahive, lead forecaster of the Weather Bureau at Logan Airport. Al did not know it yet, but he would preside over some of the most notorious New England storms ever recorded during his tenure. He had graduated from Boston College in 1937 with a degree in physics and started as a forecaster shortly thereafter, which means he had a front-row seat to the Hurricane of '38 and a destructive hurricane in 1944. He did not know it yet, but Hurricane Carol was on the way a year after the Worcester tornado.

So this man had seen a lot and had a lot more coming during his forty-one-year tenure. On this day, here was the pressing question: How to phrase a forecast that could include tornadoes to the public?

The decision? Not to mention it. Al and the team deemed telling everyone a tornado was possible would be "unnecessarily alarming" and instead took a different route. The Bureau issued the first-ever statement of "severe weather" in their bulletin at 11:30 a.m.

Officially, it looked like this:

"Windy, partly cloudy, hot and humid with thunderstorms, *some locally severe*, developing this afternoon.

To someone reading this decades later, it sounds extremely benign. We are now used to days of advance warning for the potential of severe weather, TV meteorologists talking about how tornadoes form and what to do in case of an emergency, severe weather outlooks several days out from the Storm Prediction Center, watches issued the day of, warnings issued for individual storms, and so on. In short, it is really difficult not to know something is possibly coming and no outlet is particularly shy about blaring it out into the world.

But in 1953, this tentative phrasing was pushing the envelope. For previous decades, official government policy was not to use the word "tornado." Plenty of forecasters still held this belief. What if they told everyone it could happen, panic ensued, people ran and drove all over the place out of fear, and the forecasters caused more issues than they prevented? Plus, the chance of a knock on credibility if they were wrong.

Here is the setup that stirred up such spirited debate in that hot and humid office at Logan Airport.

The surface map of the day would immediately stick out to a current-day forecaster, as it did even then in 1953. A warm front was draped across the region, with towns and cities throughout most of Massachusetts on the warm side of it. The day's high would reach 85°F in Boston and 80°F in Worcester with dewpoints in the sixties—steamy by early June standards.

Warm fronts are critical because they are often the focus for rotating storms. Winds can veer in the vicinity of these warm fronts, turning to more of a southeasterly direction at the surface instead of southwesterly. This, in turn, produces wind shear as the winds change direction with height. And wind shear is a key ingredient for tornadoes. If the air is unstable enough and it encounters an area with favorable shear, then a rotating updraft can develop. If it is strong enough, you've got your tornado.

At the same time, a trough at the upper levels was digging down toward New England. We can probably infer, given the large ridge over the central US at the time and the resulting storms here in southern New England, that there was an EML (elevated mixed layer) in place. These are air masses that originate over the higher terrain of the southwest and Mexico and can move all the way across the country to New England. EMLs are notorious for bringing about some of our most vicious supercell storms as they put a "cap" on thunderstorms until the air warms enough. Then once the cap breaks, severe weather explodes. It is estimated that this storm reached a height of seventy thousand feet.

The supercell that would become infamous had already produced monster hailstones to the west as its updraft thrust water and ice miles up into the atmosphere. Hail the size of baseballs was reported in Northfield, Massachusetts, which is about as big as it gets in southern New England. This fierce, rotating updraft gave birth to the tornado.

The first funnel touched down in Petersham, a small town near the northern tip of the Quabbin Reservoir, shortly before 4:30 p.m. Fishermen on the water later confirmed the appearance of multiple funnels, before coalescing into one massive mile-wide funnel down the road. Eyewitnesses called what they saw resembling a "huge cone of black smoke." This was not even a remotely typical sight for New England.

From there, the first lives were claimed in Barre, the next town to the east. Two more souls were taken by the storm in neighboring Rutland, and eleven more in Holden where the Brentwood subdivision was wiped out. In the Winthrop Oaks neighborhood, forty-eight of the fifty homes were destroyed.

It was at this point that the tornado grew to monstrous proportions and intensified to something truly violent. The exact period of peak ferocity occurred upon entering the highly populated and settled city of Worcester. In another sad twist of fate, it was near the city's historical population peak.

In the northern section of the city, the tornado was first sighted at 5:08 p.m. on Brattle Street. For the next fourteen minutes, chaos reigned in Worcester. Assumption College was left in such a state of ruin that the campus was eventually relocated a few miles away in 1956. Even multi-story brick buildings were torn down. A priest and two nuns were killed.

Eloise Jewers Brandt, age sixteen, was on the way home from school on a bus and stopped at Curtis Apartments in the Great Brook Valley section of the city. She later described the gruesome scene around them.

"Strewn around the street and neighboring field were bodies, dead and wounded, and blood everywhere. There was such moaning and screaming and wailing as I hope never to hear again in my life. I next remember arriving at my home in Boylston, a distance of several miles, and to this day I cannot remember how I got there."

The blueprints for the apartment complex, snatched by the ferocious winds, were later found seventy-five miles away in Duxbury, Massachusetts.

Just east of the Curtis Apartments in the Burncoat Hill neighborhood, several subdivisions were wiped clean. Sixty people were killed in a matter of minutes.

A young Carol McCrohon was enjoying her tenth birthday party when the storm arrived. In 2013, she told the *Worcester Telegram & Gazette* that she recalled seeing her family car fly by her home's second-floor window. The car was reportedly later found in Boylston, the next town over.

From there, the storm twisted its way across Lake Quinsigamond and into Shrewsbury, then onward into Westboro and Southboro. The Fayville Post Office in Southboro took a direct hit and collapsed with deadly consequences, including a baby boy. In all, twenty-one more people were killed among these three towns.

While these intense and life-changing moments were unfolding, no one at the Weather Bureau knew that a tornado was on the ground. The verification came from a surprising spot—Blue Hill Observatory, dozens of miles to the east. And the report came from someone you could call the godfather of New England meteorology.

Dr. Charles Franklin Brooks had picked up his PhD in meteorology at Harvard, and in doing so became only the second person in the country to have such a degree. He co-founded the American Meteorological Society in 1919, was the Director of Blue Hill Observatory from 1931 to 1958, and taught geography at Clarke University in Worcester. Now, he and fellow observer John Conover were witnessing the city's destruction play out from dozens of miles away on a perch 635 feet above the sea.

After 5 p.m., a call from the observatory was made to report shingles, branches, papers, and other assorted debris falling from the sky. Even before the days of technology, this was a sign that a major tornado had struck and was sending homes and lives skyward . . . eventually falling downwind.

In the end, debris fell more than one hundred miles away. Parts of books fell on Outer Cape Cod in the towns of Eastham, Chatham, and Provincetown. A mattress, thrown so high into the air that it froze solid, was found floating in the bay along the coast of Weymouth.

This on-the-ground report led to another first—the first ever "Tornado Warning" in New England. The Weather Bureau's Lead Forecaster on duty at the time, Mr. Drebert, sent it out via teletype at 5:45 p.m.

"Caution advised on severe thunderstorms with *isolated tornado activity* in the Boston area between 6 and 8 p.m. this evening."

While this was certainly some decisive and unprecedented action, it was too late. The tornado had already lifted by the time the warning was issued.

The same storm ended up producing another tornado in Sutton, Massachusetts, which tracked twenty-six miles to Mansfield, Massachusetts. Yet another twister, an F3, touched down in Exeter, New Hampshire. And a short-lived F1 tracked through Rollinsford, New Hampshire.

While not being used in the operational sense that it is today, a nearby radar was watching the storm. A research radar at the Massachusetts Institute of Technology (MIT) snapped an image of the now classic "hook echo" associated with storms producing strong tornadoes. At the time, it was just the fourth time such an image had been seen by radar.

Some controversy followed in the years ahead. The Fujita scale, a way of rating tornadoes based on damage left behind, had not been developed yet. But when it was in 1971, the Worcester tornado was retroactively given a F4 rating, the second highest. Considering the magnitude of the damage, particularly the damage found in the Uncatena–Great Brook Valley area of northern Worcester, many felt that the tornado was actually a F5.

The National Weather Service did indeed take another look in the summer of 2005 to possibly re-rate the storm. It is the only time the NWS has done such a thing for a tornado. But in the end, they could not for certain say how sturdily built the destroyed homes had been and how well anchored they were. The rating was left at F4 status with winds between 207 miles per hour and 260 miles per hour.

Today, it would likely be a billion-dollar disaster. In 1953, the price tag came in at $51 million, which remained the costliest tornado in US history until a massive tornado tore apart Wichita Falls, Texas, during the "Terrible Tuesday" outbreak of 1979.

While the casualties and costs were staggering, there was a positive in the aftermath of the Worcester tornado (in conjunction with the devastating twisters in Beecher, Michigan, and Waco, Texas, that year). Immediately following, there was a rapid reorganization of severe weather forecasting. The Severe Weather Unit in Washington, D.C., was renamed the Severe Local Storm Warning Center, or SELS. This was the precursor to the current-day Storm Prediction Center, tasked with trying to keep the loss of life suffered in 1953 from happening again.

The Windsor Locks Tornado, 1979

"My arms stretched out to reach the unseen children, an attempt to reassure them that the electricity was simply off, nothing more serious than that. Then I heard the shattering, tinkling of glass breaking and knew my living room window had blown in. That was the precise moment of realization that something awesome was happening."
—SUE BANKS, POQUONOCK RESIDENT

SUBTLETY IS OFTEN NOT A FEATURE OF SEVERE WEATHER, PARTICULARLY in this day and age. You'd be hard-pressed to go blissfully unaware of impending storms with the constant bombardment of social media and a multitude of news sources. In a way, some of the element of surprise that makes historic meteorological events memorable has been taken away from us as we wait days on end for an advertised "big ticket" weather day to arrive and play out.

Mother Nature still has her secrets, though. The autumn of 1979 decided to offer up an unexpected day that would indeed become quite memorable. In fact, it would end as one of the most destructive days in Connecticut history. There were not many who saw the Windsor Locks tornado coming. When it arrived, it struck with force.

Tornadoes hit the US more than any other country in the world, with over 1,200 snaking their way down from ominous rotating storm clouds every year. Connecticut, however, only endures an annual average of two. They're usually not the type you see in movies. Such violent tornadoes are possible but certainly not frequent in New England. Instead, the style here leans more toward small spin-ups with winds near or less than one hundred miles per hour. They come and go quickly while leaving pockets of localized damage. A fair number of them are so small and rain-wrapped that no one even can see them coming. In contrast, the Worcester tornado in 1953 is one of the few classic tornadoes to touch down in the region. They are the exception rather than the rule and rare is the day when forecasters are truly worried about a devastating twister spinning up.

If you were going to see one rip across the landscape, there is a natural window of choice that exists from late spring through summer. Over 80% of all New England tornadoes occur during this time with the highest risk in July. In other words, we do not think of starting up the storm-chasing van during a leaf-peeping trip. By early October the seasons are shifting gears and so are the expectations. A strong autumn gale? Par for the course. A hurricane roaring up the coastline? No one would be surprised. But a significant tornado twisting a path of destruction is quite out of character for the time of year. Since 1950, only a handful of EF4 or EF5 tornadoes have been recorded across the entire country during the month of October. It is the least likely month of the year for a violent tornado to form.

Throwing aside these mitigating factors, October 3, 1979, frankly just did not look like the sort of day you'd expect historic weather to play out. Few observers of the atmosphere would peer at the maps that morning and have their hair stand up.

When seeking out the glaringly obvious severe weather days in New England, a key variable to look for is something called an elevated mixed layer, or EML. These air masses move in from the High Plains or Southwestern US and can drift overhead in a westerly flow aloft. Essentially, you end up with an atmospheric profile where, once the surface is heated, air will rise rapidly, which leads to explosive storm development. Prime and well-known examples of this process at work are the Worcester tornado in 1953, a F4 that struck Hamden, Connecticut, on July 10, 1989, and the Springfield, Massachusetts, tornado on June 1, 2011. The playbook calls for southwest to northwest flow aloft and extreme instability above the ground. Then watch out below.

But on this October day, no such profile existed. It was actually rather cool and gloomy across northern Connecticut with low clouds and drizzle to go with the morning coffee. One of those sleepy autumn mornings where thoughts drift toward the gathering cool season and away from the rolling thunder of summertime. The previous evening had featured some strong thunderstorm activity in Pennsylvania and New Jersey, and heavy rain was ongoing to the southwest of Connecticut at daybreak. An 8 a.m. satellite image showed an impressive comma-head shape to the clouds as the squall line pressed toward the Atlantic. Southern New England remained entrenched in a cool easterly flow, north of a warm front stationed along Long Island Sound. The feel would not be particularly alarming to anyone venturing out in the mist. No warmth, no castellanus clouds showing the day's hand.

Up above the gloom, ingredients were conspiring for trouble. A weakening upper-low drifting over from the Great Lakes was bringing along a pocket of cooler air aloft, which can help to destabilize the atmosphere. Initially a closed low at five hundred millibars, it was opening up and triggering more thunderstorms as the upper-level disturbance traveled into the northeast.

Out ahead of it, a southerly flow was dragging mild air up the eastern seaboard. A few thousand feet off the ground, the atmosphere was warming up as this tongue of mild and saturated air was drawn into southern New England. Helping to urge this air along was an increasing intensity of wind, seen on weather balloon (radiosonde) soundings that day.

And so upstairs, the pot was starting to boil. Where people live and do business the picture was subtle but progressing. A primary area of low pressure had moved into western New York with another weak area of low pressure trekking toward Connecticut from eastern Pennsylvania. It was no powerhouse, but sometimes severe weather is all about the small, mesoscale details. By lunchtime this small wave of low pressure had traveled into the Lower Hudson Valley and by early afternoon into western Connecticut.

Out ahead of this mesolow (low-pressure center) the individual cell that would produce the most damaging tornado in Connecticut history was alive and well but had not achieved full potential yet. South of Long Island at midday, it reached New Haven at 2 p.m. and was heading right up the Connecticut River Valley. Here again, we find a very unusual circumstance. Violent southern New England tornadoes generally have some west-east component to their movement and typically are not tracking south-north. A unique event had nearly come to fruition.

Still, any observer in northern Connecticut would not be expecting a tornado to arrive at any moment. Winds were still east-northeasterly and temperatures cool in the sixties. And it wasn't just the laymen in the dark. There were no severe thunderstorm or tornado watches in place, and the National Weather Service issued no tornado warning before touchdown.

At 2:50 p.m. a gusty northeast wind was still blowing at Bradley International Airport, which also happened to be home to the National Weather Service. Little did they know they were at the conjunction of all the small meteorological features that would come together in this ill-fated spot. The cold pocket of air ten thousand feet up had arrived. A ribbon of warm and moist air had traveled right up the Connecticut River Valley along with the storm. The layering of these two would help produce extreme instability and rising motion. To tie it all together, a mesolow had arrived right on the "triple point" where a warm front, cold front, and occluded front all intersect to maximize lift and wind shear. In a day where everyone else just received heavy rain, all the ingredients were perfectly aligned in this one spot.

Reports of the first touchdown came from Poquonock, a village on the northern edge of Windsor, Connecticut. A funnel dropped from the clouds just south of the Farmington River and used Poquonock Avenue (Route 75) as a highway north. Torrential rainfall veiled the vortex from sight, but damage began in earnest. The starting point was Poquonock Elementary School, where, by a stroke of good luck, students had been let out early for the day. No one was hurt even though the roof was largely destroyed.

Just one hundred yards up on the other side of Route 75 sat the Poquonock Community Church, which suffered a major blow as the tornado began to widen. The roof and steeple of the 255-year-old parish were ripped off, but amazingly all but one of the art deco windows were left intact (to be included in a new sanctuary during the rebuild). Carmon Funeral Home directly across the road was also damaged. Quickly, the situation began to escalate.

Buildings were the first victims but when it came to the greatest human suffering, ground zero was to occur just up the road in the Settler Circle and Pioneer Drive neighborhood. Now spinning violently and expanding to its greatest width of 1,400 yards, the tornado intensified with raging winds over two hundred miles per hour that engulfed the small community. Homes were wiped clean off their foundations, a hallmark of only the strongest tornadoes.

And it was here that the storm's only deaths occurred. Two construction workers who had been at the Hartford National Bank were hit by flying debris. William Kowalsky, a twenty-four-year-old man with a wife and young daughter was killed instantly when lumber was launched through the truck he had taken cover in. A co-worker who also had taken refuge in the truck succumbed to his injuries weeks later. The bank they had been working on was wiped out, and checks from Windsor would later rain down on North Adams, Massachusetts, in the Berkshires.

On Settler Circle, just east of Route 75, a forty-two-year-old woman named Carole Dembkoski home alone while her husband and two sons were in neighboring Bloomfield. She was grabbed by the tornado and thrown over one hundred feet from her home. Dembkoski was likely killed on impact, but rescuers would not locate her body under piles of debris until the following day. Nearly five hundred others were injured.

Miraculously, there would be no more loss of life on the journey north. The same could not be said about property damage. Continuing up Route 75, a number of businesses including the Koala Inn, Frank's Diner, Bradley Cleaners, Rice Hardware, and a number of car washes and gas stations were all partially or totally destroyed. Over one hundred buildings in all would be reduced to rubble along its path. Gravestones at St. Joseph's Cemetery were toppled. Still, the target that was about to take the biggest hit was just ahead—Bradley Airport.

Aside from being the main airport hub for Connecticut and western Massachusetts, Bradley Airport happened to be home to one of the greatest aircraft collections in the country. The Bradley Air Museum sat just west of Route 75 and held approximately 120 historic aircraft from bombers and transports to helicopters. Indiscriminately, the tornado began tossing them like a child's toy planes.

The first to fly wasn't at the museum itself, but outside the Ramada Inn next door. An F-102 Delta Dagger fighter plane that had been on display was flipped and tossed into the road. The museum ahead was just several hundred yards east of the National Weather Service office at Bradley's main terminal, where a 90-mile-per-hour gust was measured. The *Hartford Courant* would later run the unflattering headline "Forecasters located right under tornado, never spotted funnel."

Quickly, the museum's historic collection was in ruins. The largest of them all, a C-133 Cargomaster weighing over sixty tons, was flipped over with both the nose and tail ripped off. Of the thirty planes sitting out on display, sixteen were destroyed and ten others significantly damaged.

Still more were inside a twenty-four-thousand-square-foot hangar, which the tornado decided to barrel directly over. Several employees who had been outside staring in awe at the unusually colored green sky ran for shelter inside the hangar and were fortunate to survive. The roof was ripped off and nearly every one of the planes inside were victims of at least minor damage. Some were hanging from the roof, others littered with debris, and still more missing propellers, wings, or tails. A short distance across the airfield sat a brand-new set of helicopters belonging to the Army National Guard, and several of the Sikorsky CH-54 Skycranes were obliterated by the swirling winds.

The severe damage of such pricey real estate catapulted the Windsor Locks tornado to one of the ten costliest in US history (a ranking it held on to until 2013). Tearing through vintage aircraft and National Guard helicopters, along with the devastation through Poquonock, came with a price tag of over $200 million at the time and over $800 million in today's dollars.

While all this mayhem was unfolding, one United Airlines flight had a front-row seat. Captain George Diehs peered into the tropical rainfall from the cockpit of his 727. Flight 220 from Chicago was about to reach the tarmac, but something did not feel right. A former helicopter pilot in the military and now a twenty-three-year veteran with United, Diehs noted that the plane was a little off course from where it should be on approach. Combined with the quickly changing wind and poor visibility, he aborted the landing just two hundred feet above the tarmac. The captain's split-second decision saved the lives of 114 passengers, some of whom spotted the tornado outside their cabin windows as Diehs slammed the throttle and aimed back into the heavens.

Damage was maximized in Windsor and Windsor Locks, but the storm was not done yet. On what would become an 11.3-mile path, it continued due north into Suffield where over two dozen tobacco barns were extensively damaged at a cost of $2 million. Then, crossing state lines, it continued to tear up trees into Southwick and the Feeding Hills area of Agawam, Massachusetts. Several more roofs were wrecked but the ferocity of the twister was beginning to diminish as it moved into a cooler and more stable air mass. It was here in Hamden County that the track came to an end.

She wasn't home at the time, but Connecticut Governor Ella Grasso just happened to live a couple blocks from the tornado's path in Windsor Locks. Upon hearing the news, the governor quickly instituted a curfew to prevent looting and activated approximately five hundred National Guardsmen. President Jimmy Carter would later declare the affected towns a federal disaster area to open up funding for recovery.

Famed tornado expert Dr. Ted Fujita, for whom the Fujita scale to rate tornado intensity is named, would come in to survey the storm's rampage. The result would be an incredibly detailed diagram indicating the path and strength of the Windsor Locks tornado. Even localized downbursts, which were prominently displayed on the right side of the tornado's track, were color coded and mapped out. The final of these downbursts was noted across the Westfield River in West Springfield and Westfield, Massachusetts.

Based on the most severe damage found in Poquonock where homes were annihilated, Fujita would rate the Windsor Locks tornado a F4 with maximum winds of 207 to 260 miles per hour. The F4 rating corresponds to "devastating damage" on the Fujita scale where "Well constructed homes are leveled, structures with weak foundations blown away, cars thrown, and large missiles generated." To be sure, this was a storm of rare ferocity unmatched in at least one hundred years. Also speaking to the unusual nature of the event, it would be another thirty-four years before there was another violent tornado *anywhere in the US* during the month of October.

The only tornado in Connecticut records that previously matched such power was the Wallingford tornado of 1878. While Windsor Locks remains the costliest, Wallingford is the most deadly with thirty-four killed. Striking during a more typical time of year, this infamous twister touched down on August 9, 1878, and leveled the town so badly that it became a morbid tourist attraction. Though they at least raised funds upon their arrival to help the survivors.

PART SIX
Ocean Storms

Portland Gale, 1898

THERE IS NO SEPARATING THE LOVE AFFAIR BETWEEN NEW ENGLAND and the ocean. Our ancestors arrived by boat, lived off the sea, and made fortunes with their vessels. Still today, we generally go to great lengths to get near its fresh breeze and the picturesque dunes that greet it. Whether paying a premium to live by the shoreline or trying to carve out a week of summer vacation along it each year, the ocean calls us. The relationship is not always mutually beneficial, however. An extreme slumbering power resides in the sea, awoken by great gusts and shifting tides. When angry, it goes where it wants and consumes what it wants. The historic gales of the Atlantic have taken their fair share of life and property.

These gales often fall in a category between two types of systems— tropical cyclones and winter storms. Some of the most infamous benefit from an injection of tropical energy or, at the very least, the still warm waters of the Atlantic in autumn. At the same time, the atmospheric circulation is undergoing its transformation to winter. The jet stream begins to return from its polar vacation up north and digs into the mid-latitudes to add a dose of wind energy to developing storms. Tapping both these elements at the transition of seasons, autumn gales can rival any other great storm in New England.

Their timing also puts many aspects of life at risk. Since it is not yet winter, they may strike early enough to tear down a prolific number of still-foliated trees. Ships not yet hunkered down or put into drydock for the cold season can be caught unprepared. Preparations along the coast for winter may not yet be finished. These storms can pose a unique hazard, which was especially risky in the heyday of American shipping and commerce at sea.

Tracking them, at least, is one of the better success stories. With the implementation of high-resolution satellite imagery, it is now much easier to spot developing storms and watch them in real time. GOES-16 and GOES-17, two of the most recent US satellites placed into orbit, look down on us with fresh scans every five to fifteen minutes, and as rapidly as one minute apart if there is a storm of particular interest to NOAA. Equipped with a lightning mapper, bursts of convection as a gale explodes over open water can be instantly monitored and relayed to ships at sea. It is not impossible to get into a life-threatening situation on the temperamental Atlantic, but it is now a lot easier to avoid.

We would hope and expect maritime disasters to become very infrequent in the future. For seaside homes and businesses, it is all about risk management. Great gales, hurricanes, and blizzards will always keep coming. The seas continue to rise as the planet warms, and coastal flooding is increasing in New England and other parts of the world. Will we continue our rocky relationship with the alluring ocean? Destructive as some of these ocean storms are, my wager would be most definitely.

The Portland Gale, 1898

"Such a violent outburst of the elements had not occurred in this vicinity within the memory of the oldest inhabitant. The wild fury of the wind and driving snow continued without abatement until late in the afternoon. At times the force and roar of the tempest were so appalling as to be indescribable."

—LIEUTENANT WORTH G. ROSS, ASSISTANT INSPECTOR
OF LIFESAVING STATIONS IN BOSTON

MANY FAMOUS STORMS DERIVE THEIR NAMES FROM THE DATE, LOCATION, or a holiday if one should strike on such an occasion. Only one of the most notable storms in New England history is known by its primary victim. The Portland Gale of 1898 destroyed 150 ships during one wild November weekend, though none grabbed the headlines nor had a higher loss of life than the steamship *S.S. Portland*. It remains one of the worst tragedies to ever occur in New England waters.

The time of year lends itself to dramatic gales. Cold air is pushing farther southward as winter begins its descent into the mid-latitudes. At the same time, the ocean is still warm and tropical energy lurks. Much like the Perfect Storm in 1991, it was a combination of these elements that was custom-made for a marine disaster and created the Portland Gale. Fast-moving and potent, it thrust the ocean into a frenzy and caught many fishermen and sailors in suddenly ferocious seas. The ill-fated steamship may be the most well-known, though not alone in the tempest. In just two days 456 people, most at sea, perished. Nearly half of them were on the *Portland* when it went down. There were no survivors when the ocean claimed it.

Just two days prior was Thanksgiving. After giving thanks for their blessings, many were preparing to travel back home. For anyone heading to Maine, the *Portland* was a popular option running a route from Boston to Portland. It was a local version of the *Titanic* before there was a *Titanic*. A luxury steamship capable of carrying nearly eight hundred passengers, it was outfitted with chandeliers, red velvet carpeting, fine china, and the extravagances of late nineteenth-century life. The run of the vessel was nearly the length of a football field, measuring 281 feet bow to stern. Crafted by the prolific New England Shipbuilding Company in Bath, Maine, it had earned a reputation as a solid and dependable ship.

While crews readied the *Portland* for a Saturday-evening departure from India Wharf in Boston, nature was conspiring against her. On the heels of Thanksgiving, a storm system was heading up into the Great Lakes. From there, a classic transfer of energy across the Appalachians to the coastline was set to begin. A strengthening area of low pressure off the coast of Virginia was the seed for what would quickly become a dangerous and powerful ocean storm, paralleling the eastern seaboard toward Cape Cod.

On Friday, November 25, the forecast for the next day called for "fair, continued cold, brisk westerly, shifting to southerly winds." For Sunday, "Unsettled, rain or snow possible in afternoon or night. The foul weather depends on the movement and development of a storm now apparently in the Gulf of Mexico."

Even in 1898, there was at least an inkling of stormy weather to come. Though the magnitude was surely underestimated and the fine details impossible to predict in this early stage of weather forecasting. By midday Saturday, as the storm drew closer, the situation became more serious in the eyes of meteorologists. The Weather Bureau issued an update—"For Maine, New Hampshire, and Vermont: heavy snow and warmer tonight. Sunday, snow and much colder. For Massachusetts, Rhode Island, and Connecticut: heavy snow tonight. Sunday, snow followed by clearing and much colder, southeasterly, shifting to northeasterly gales tonight, and northwesterly gales by Sunday."

Underway was the classic rapid intensification of a nor'easter, or bombogenesis. This occurs when then sea-level pressure falls at least twenty-four millibars in twenty-four hours, taking a run-of-the-mill storm to new heights in less than a day. Strong winds are the result of such a quick pressure drop as air rushes into the center of a storm and rises rapidly, evacuating out at the upper levels. The Portland Gale was ready to howl.

At the helm of the *Portland* was Captain Hollis H. Blanchard, a native of Belfast, Maine. He had sixty-three crew members along with him and certainly knew that a storm was brewing. The captain had made a trip to the Boston Weather Bureau's office that day. Even without that knowledge, a man of the sea would have recognized the growing signs of an incoming storm. The intel from meteorologists, plus the dead calm that was initially experienced, were harbingers that caution was likely the best path forward. Though perhaps he had a sense of false confidence or had not gained enough familiarity with the ship yet. He had been put in command of the *Portland* only three weeks earlier.

Since everyone went down with the ship, an element of mystery remains. To start with, it is unknown exactly how many were onboard or who they all were. The manifest was only kept aboard on the *Portland*, so it is impossible to say with certainty how many got underway from Boston. It is estimated that there were approximately 130 travelers along with the captain and crew for a total of 192. It is because of the *Portland*'s tragic fate that manifests are now left onshore.

As for why the ship even left with a ferocious storm gathering? Some accounts maintain that Blanchard set sail against orders to stay in Boston for want of getting back home to see family. Others suggest the shipping line itself told Blanchard to get underway to avoid delays with anxious holiday travelers. His son made comments afterwards to this effect. We will never know for sure why he took a ship meant more for the purpose of moving lots of people than battling high seas out into the Atlantic. But we do know that the *Portland* left India Wharf at its usual 7 p.m. on the evening of Saturday, November 26. With visions of Casco Bay ahead, it was immediately under attack from an angry ocean.

On the way out of the harbor, the steamships *Kennebec* and *Mt. Desert* both reported watching the *Portland* glide by. Captain Jamie Collins of the *Kennebec* had started the journey out of the harbor's safety into the

Atlantic, but turned around upon seeing the conditions. It is said that he blew the ship's whistle to warn the *Portland* but received no response. At the mouth of Boston Harbor, the keeper of Boston Light at Deer Island also got a glance of the *Portland* on its way into the inky darkness of an intensifying gale.

Upon reaching the waters along Cape Ann, just north of Boston, the *Portland* was engulfed in a full-scale blizzard. Snow was flying, seas were building to thirty feet, and winds were roaring. Hurricane-force winds were observed to the south on Nantucket, with a peak gust to ninety miles per hour at Blue Hill Observatory (elevation 635 feet) outside of Boston. Weather-reporting stations were less numerous in the 1800s but it can be extrapolated that gusts over the open waters off Cape Ann were likely in excess of eighty miles per hour.

Four hours after leaving the pier in Boston, the *Portland* was sighted near Thacher Island, which sits roughly a half mile off the coast of Rockport and features twin lighthouses to warn mariners of dangerous waters. Several schooners reported seeing her flailing in the turbulent ocean, including the *Grayling*. Less than an hour later, at 11:45 p.m., the schooner *Edgar Randall* almost ran straight into the steamship. At this time, it was reported that the *Portland* no longer had operating lights and was in rough shape struggling to make headway. Fighting a losing battle against the wind-whipped waves, it was now moving southeast toward Cape Cod instead of northward to its intended destination of Maine. It is likely that Captain Blanchard, realizing a fatal error, was attempting to get back out to open ocean to ride out the storm at sea instead of tracing the dangerous shoals near the coast.

This logic may be backed up by a report from Provincetown the next morning. Lifesavers at Race Point heard the four shrill whistle bursts of a troubled steamship at 5:45 a.m. Before a rescue effort could be mustered, the whistles went silent and were not heard again. That evening, a patrolman found a white life jacket that turned out to be from the *Portland*. Far more gruesome things were about to wash ashore in the days to come.

Bodies began to reach the beaches of Cape Cod that night as the tide rose. All the passengers wearing watches had the time stopped at 9:15, indicating the demise of the ship was likely mid-morning of November

27. We will never know for certain exactly how the end came, but when the wreckage was later found with the destroyed schooner *Addie E. Snow* less than a quarter mile away, the general feeling was that they collided and sank together into the deep.

Only one passenger was alive when he reached dry land. According to historian Al Snow, one of the *Portland*'s quartermasters was found by a surfman patrolling from the Nauset Lifesaving Station. When the surfman turned the body over, the survivor gasped "For God's sake—do not leave me!" He died before the surfman could return with help. It must be said that the brave crews manning lifesaving stations from the South Shore to Cape Cod did save at least 120 people during the storm. It was terrifying work and took considerable courage to accept the responsibility, with many enlisted in the Lifesaving Service becoming victims for their efforts.

Over the next three days, thirty-four bodies from the *Portland* would wash up. The majority collected near Race Point in Provincetown but wreckage reached as far south as Chatham, including the six-foot double steering wheel recovered at Nauset Inlet (later destroyed by the *Portland*'s owners). The *Portland* was not alone in its fate. Another one hundred bodies washed up along the South Shore on the inside of Massachusetts Bay from Hull to Plymouth. It was pure catastrophe for the maritime community. Sadly, it was more confirmation of the "ocean graveyard" label given to the waters that line Outer Cape Cod. Just between Wellfleet and Truro, there are an estimated one thousand shipwrecks lying on the ocean floor. Rapidly intensifying storms like the Portland Gale, along with the challenging shoals and shape of the coastline, produce frequent conditions with these tragic endings.

With communication and railroad travel greatly disrupted, news of what had happened took a while to reach Boston. A newspaper reporter stationed in Hyannis was seeing and hearing of the catastrophic loss at sea by way of telegram from Truro, but in order to file his story he had to make it back to the city. The journey required a combination of horseback riding, walking, and finally a train. It would not be until Tuesday, two days after the *Portland* sank, that word got out to the masses.

Nearly a century would pass before the exact resting spot of the *Portland* could be confirmed. Over the years fisherman occasionally found pieces of it when pulling up their nets and oceanic detectives followed the trail of clues to search. Rumor had it that the wreck had been found on the bottom several miles north of Cape Cod, but firm evidence could not back it up. Then in 1989, the Historical Maritime Group of New England took up the cause (just four years after the *Titanic* was located in the North Atlantic). Looking back at the reported sightings during the *Portland*'s final hours and taking into account all the watches that stopped at 9:15 a.m., they went to work sleuthing its whereabouts. By tracking the way debris drifts in the area and local currents, they used a side scan sonar to search a smaller section of Massachusetts Bay. The group did indeed locate a signature on the bottom that had characteristics of a steamship such as the *Portland*, but technology of the late 1980s did not quite have the resolution to definitively say it was the ship.

In 2002, the Historical Maritime Group was proven correct. A team of researchers from the University of Connecticut teamed up with NOAA to use sonar and deploy a remotely operated vehicle, or ROV, to the site. The resulting video confirmed that the *Portland* lay three hundred feet down on the edge of Stellwagen Bank, about twenty miles southeast of Gloucester and to the north-northwest of Provincetown. Its bow was found facing southeast, which would suggest the ship was broadside to the waves before rapidly sinking. Three years later in 2005, the wreck was added to the National Register of Historic Places.

The demise of the steamship is the largest by scale during the storm, though it is certainly just a chapter in the Portland Gale story. With nearly 150 other ships destroyed at sea, news of just how disastrous the weekend was slowly coming back to Boston and beyond. In the busy fishing hub of Provincetown, nine wharves were obliterated, striking a severe blow to the industry and lifeline of the town. Dozens of Cape Cod ships were lost. It was a crushing event in what would be the most destructive year in the history of New England shipwrecks. The Portland Gale was responsible for 70% of all vessels lost.

Along the South Shore a surge of up to ten feet in Cohasset had pummeled the coast. Sheer force of the sea changed the local geography in just twenty-four hours. Before its arrival, the Humarock peninsula ran from Scituate's Third Cliff down past the Fourth Cliff to the mouth of the North River. The coastline from Scituate to Marshfield features a rolling set of cliffs, which are actually drumlins left over from the most recent ice age. Intense storm surge created a new cut, destroying the land that had connected the third and fourth cliffs. With this redirection, the North River broke through a different location and earned three more miles of length! As the coastline rearranged, Humarock briefly became an island before reconnecting. Scituate's loss was another's gain, as the spit of sand moved south of the river and attached to Marshfield instead. In the following years the mouth of the North River continued to widen and has become vastly larger than in 1898.

The entire south shore region between Boston and Plymouth was a mess of downed electric and telegraph poles. Numerous houses were damaged or destroyed by the strong winds and surf, sending some chimneys tumbling down and shattering windows. The seawalls erected to act as a defense against the waves were left in tatters, particularly so near Brant Rock in Marshfield, which juts out to face the northeast swell.

Typical debris from a large coastal storm features seaweed, lumber, and lobster traps. With the devastation taking place at sea during the Portland Gale, there was a considerably larger variety of goods and materials washing up. Furniture, clothing, boats, and stoves were strewn about. In Hull along the oceanfront stretch of Nantasket, a Ferris wheel and roller coaster were destroyed.

When reading through the summaries of the Portland Gale and its wrath, understandably a lot of the focus is put on the tremendous toll it took at sea and on the immediate coastline. If you put that aspect of the nor'easter aside, then quite a bit could still be made of the large snowstorm that was pounding the rest of the region at a very early point in the season. The cold that poured in after Thanksgiving and the snow piling up during the storm were impressive to say the least!

It was an all-out blizzard with blinding snow covering the landscape from New York City to Maine; many locations picking up over a foot of snow. Even the coastline participated with Boston receiving 12.5 inches. If your interests were not at sea, the enduring tale of the weekend gale was all about the raging snowfall shocking people for its ferocity before winter even had settled in. Roads, trolley, and railroad lines were blocked and shut down. With records blown away, Connecticut was in a complete standstill. New Britain's Central Railway & Electric Company declared the storm "the most severe storm" they had ever had to deal with.

One surviving diary from the time belonged to a sixteen-year-old young woman named Mary Noyes, who lived in Stonington. Coming off the Thanksgiving holiday, she felt compelled to describe the sights of the storm.

On the morning of Sunday, November 27, she noted a "very stormy night, the wind blew terrible all night and this morning a dreadful blizzard is raging, the roads are banked very high with snow." And on the following morning "such high banks! Mr. Lord began to clear the roads but did not get but a little way in all day."

Another snowfall trailed the gale on November 30, and then Miss Noyes recapped the extreme stretch on December 3 after spending a time attempting to travel outside the home. "I can't say but that I have taken one wild ride. There was hardly a place on the road where a team (of horses) could pass one another. The snowbanks were for a long distance a foot above the wagon wheel."

A stretch of significant winter chill had swung in behind the gale, and so the deep-winter scenes Noyes described were not surprising. Little melting gave relief for the cleanup efforts with highs just in the twenties and thirties through the rest of the month.

Ten years previous, New York City had taken the full brunt of the "Great White Hurricane." The gale did not measure up to that March beast, but it was another record setter. The ten inches that fell during the Portland Gale makes it the largest snowstorm ever recorded so early in the season, and it is a mark that still stands today. The same can be said for Boston and Worcester, Massachusetts (eighteen inches). In Concord, New Hampshire, the sixteen inches stood as the record holder until demoted to second place in the freak October snowstorm of 2011.

The Perfect Storm

"Meteorologists see perfection in strange things."
—Sebastian Junger

Many big storms live in our memories and occasionally the notable ones may show up on local TV during their anniversary. We chat about them as benchmarks. The classic "I remember where I was when..." events. But not many storms garner enough attention to make the silver screen for a larger audience. In late October of 1991, one such storm brought together so many unusual ingredients and dramatic storylines that it climbed into the realm of pop culture around the world.

The meteorology behind "The Perfect Storm" is interesting enough, but what really catapulted it into the spotlight was, unfortunately, human tragedy. Trapped in the heart of a raging ocean, the Gloucester, Massachusetts, fishing vessel *Andrea Gail* sank, taking all six crew members down with it. The story of its battle on the high seas and eventual demise became inspiration for Sebastian Junger's *The Perfect Storm* and then a major motion picture starring George Clooney as Captain Billy Tyne. But how could something so big creep up on seasoned fishermen, resulting in loss of life? In short, the mixing of seasons stirred up something quite unexpected.

By mid-autumn in New England, the tranquil and sunny times of September are slipping into the grasp of an incoming cold season. Days quickly grow darker while buttoning up and hunkering down gets underway. Patio furniture migrates to its winter home in the garage. Boat owners eye the water warily, weighing whether it is time to haul in for the season or hold on for a few more runs.

Right at this transition point, the first big gales and nor'easters begin to stir. While they awaken, the Atlantic hurricane season is not quite ready to give up. Ocean temperatures are still warm, and the intermingling of coastal storms and energetic tropical storms can breed wild weather. "The Perfect Storm" brought all these ingredients together over multiple days for one bizarre event that brought widespread damage along the coastline from Florida to New England.

Fierce gales with towering waves and strong winds are nothing new for the North Atlantic. What made this one special was not necessarily its intensity, but the way it came together and played out over several days. The tempest was a product of interaction between several separate weather systems. From an autumn front to a deepening storm, then ending its life as an unnamed hurricane. Indeed an atmospheric script made for Hollywood.

Also known as the "Halloween Nor'easter" or "No Name Storm," credit for the most frequently used moniker "Perfect Storm" goes to National Weather Service Meteorologist Bob Case. The name wasn't meant to have a positive connotation. Instead, more of a "you couldn't possibly have everything go worse" meaning more akin to Murphy's Law. It was a constant game of catch-up for forecasters and those at sea to keep pace with the rapidly changing conditions that unfolded.

The first element to be set in motion was a potent cold front pushing off the coast on October 27. A typical occurrence, however once offshore the front stalled and an area of low pressure formed along it. This is where things immediately started to get dicey.

The newly born storm system now swirling offshore would, in most circumstances, start to head out to sea. In this situation, a massive area of high pressure came down across eastern Canada and blocked that from happening (something we see a lot of in this book, and indeed a "good" way to produce unusual and high-impact weather). What made this high rather unusual is that its strength resembled something expected more in mid-winter than October, with a sea-level pressure of 1,046 millibars.

A rule of thumb in weather is "forecast the high, forecast the storm" and in this case it was certainly a large part of the story. The high-pressure ridge would end up capturing the storm and drawing it westward over time, which meteorologists call retrograding. Most storms move to the northeast or east when exiting the coast, but this one would drift backwards toward the shore.

On top of the unusual motion, very high pressure helped to produce a tremendous pressure gradient, or rate of pressure change over a distance. Think of a weather map you may see on TV or online showing all those lines circled around a storm. Those are isobars, or lines of equal pressure. The more there are, the larger the pressure gradient and most importantly, the stronger the wind you can expect. The Perfect Storm packed in a ton of them, with a 1,046 millibars high smashed up against what would become a 972 millibars low by the morning of October 30. This was the gale's peak intensity, centered 340 miles south of Halifax, Nova Scotia. That difference of eighty-four millibars meant winds topping seventy miles per hour on the coast of Massachusetts and roaring at hurricane-force over the open ocean.

This process played out rapidly, so that on the morning of October 28 NOAA's Ocean Prediction Center released a bulletin warning of a "dangerous storm" centered south of Nova Scotia and east of Nantucket, Massachusetts. By then, the Atlantic was already a frenzy of spray and waves tall as city buildings. A weather buoy in the open water south of Nova Scotia recorded a wave reaching one hundred feet high! Waves of ten to thirty feet were pounding the eastern seaboard from Maine to North Carolina at this time, chewing up dunes and beachfront property.

It was on this same day that the *Andrea Gail* was lost. The seventy-two-foot swordfish boat was located approximately 150 miles east of Sable Island and had likely been steaming home from the Grand Banks. Experienced as the *Andrea Gail* and her crew were, something catastrophic happened on the water that no one survived to describe. No distress call was made, the boat's radio failed, and after an exhaustive ten-day search nothing was found by the Coast Guard. Oddly, their Emergency Position-Indicating Radio Beacon (EPIRB) did eventually wash up on Sable Island on November 5 with its switch in the off position.

The last person to speak with Captain Billy Tyne was Linda Greenlaw, another fishing captain at the helm of the *Hannah Boden*. She later recalled their last conversation, with Tyne saying "The weather sucks. You probably won't be fishing tomorrow night." Typical chatter for a couple of Gloucester fishing comrades. No one knows why, exactly, the *Andrea Gail* sank. The entire crew, which in addition to Tyne included David Sullivan, Bob Shatford, Dale Murphy, Alfred Pierrre, and Michael Moran, were swallowed up by an indifferent ocean.

Being at sea was undeniably treacherous, but up and down the coast relentless waves were hazardous to anyone who ventured too close. In Narragansett, Rhode Island, a fisherman was swept off the rocks by waves and killed. Another fishing in New York was blown off a bridge and died. Many, however, heeded warnings that meant the biggest danger was to coastal property and the land it sat on.

Especially hazardous was the long duration nature of the storm. High surf was rolling in over multiple days and multiple high tide cycles, each one successively higher as water piled up and was unable to escape between tides. A NOAA report after the storm summarized the extensive issues up and down the eastern seaboard.

> *"Beach erosion and coastal flooding was severe and widespread, even causing damage to lighthouses. Hundreds of homes and businesses were either knocked from their foundations or simply disappeared. Sea walls, boardwalks, bulkheads, and piers were reduced to rubble over a wide area. Numerous small boats were sunk at their berths and thousands of lobster traps were destroyed. Flooding was extensive invading homes and closing roads and airports."*

Some of this was due to the fact there was not just one, but two systems were churning up the water. To become the "Perfect Storm" a merger between them would be required.

On the same day (October 27) that the nor'easter had formed, a new hurricane named Hurricane Grace had also joined in. On top of all this meteorological mess, Grace itself was born out of a previous subtropical disturbance. It was a very busy time for any forecaster with interests in the Atlantic. Grace strengthened and peaked as a Category 2 hurricane with 105 miles per hour sustained winds in the warm waters south of Bermuda. Piling on, it contributed by sending high swells toward the US.

Slowly it took a path to the northwest, eventually making a sharp turn to the east and passing fifty miles offshore of Bermuda on October 29.

From here, it got weirder. The hurricane was "sucked up" and absorbed by the nor'easter raging south of Nova Scotia and quickly lost its tropical characteristics. Grace likely brought along an injection of energy into the original storm by adding moisture and heat. As it mixed in the new and improved low-pressure system began to drift southwest back toward the coast through October 30 and 31. The combination of the two systems made for water levels rarely seen on the east coast and were the primary reason so much destructive power was realized.

Peak gusts were recorded as seventy-nine miles per hour in Provincetown, Massachusetts; seventy-eight miles per hour in Chatham, Massachusetts; and seventy-four miles per hour on Thatcher Island off the tip of Gloucester, Massachusetts. Impressive, but nothing that would compete with the strongest blizzards or hurricanes. However, there is a lot that can be attained through longevity. For instance, Chatham's peak gust was notable but more importantly the site recorded gusts over sixty miles per hour for fifteen straight hours. You do not need all-time wind for all-time flooding, which was illustrated by the Perfect Storm.

Days of sustained winds of thirty to fifty miles per hour piled up water along the coast. Not only were the winds persistent, but also uniform over a large patch of ocean. This is called "fetch," the distance over water that wind blows from the same direction. Brief wild winds can whip the ocean into a froth and bring in a quick surge, but a substantial fetch will generate long-period swells traveling long distances, resulting in some of the largest and most destructive waves a storm can offer. In this instance, ten- to thirty-foot waves came crashing ashore from the Canadian Maritimes to the Outer Banks of North Carolina.

In New Jersey, water heights reached levels not seen in forty-seven years since the Great Atlantic Hurricane of 1944. Coastal locations in Delaware, Maryland, and Virginia saw the highest water in decades. Ocean City, Maryland, set a record (at the time) with a high tide of 7.8 feet, besting the previous mark from a March 1962 storm. Since a barrage of waves kept coming in, outgoing water from large rivers was backed up. There were reports of flooding as far inland as Washington, D.C., on the Potomac River and Albany, New York, on the Hudson River! Coastal flooding was even seen all the way down into the Bahamas and Puerto Rico. Though it was in Massachusetts that the worst destruction and watery wrath was found.

Not since the Blizzard of '78 was the ocean so high in Boston. Tides came in a full four feet above normal in the city. A home in Scituate was thrown off its foundation by the raw power roaring ashore. Another 118 were destroyed and 453 suffered major damage in that town alone. On the outstretched arm of Cape Cod, water even managed to surpass '78. The Atlantic ran over Coast Guard beach in Eastham like a freight train. Lofty dunes were reduced to nothing as they gave way to the surge of an angry ocean.

To the north, a huge one-hundred-foot breach opened up in the towering dunes behind Ballston Beach in Truro. Water rushed in and blew out nearly two hundred yards of marsh behind what had been a protective barrier. The ocean would make it all the way to Route 6 before stopping.

The cut washed out by the Perfect Storm would go on to be a long-lasting weak spot in the sandy fence of the Outer Cape, breaching and eroding repeatedly ever since. If you climb the bluff today, you'll find a photo from October of 1991 to compare how much has been lost since the storm. It features a quote from local resident Andy Shultz who recalls the scene as "incredible, the water just smashed through like the dune wasn't even substantial. It was like a wall of water you could see even in dim light."

Chatham suffered a great loss of property with six homes along Pleasant Bay destroyed and thirty others severely damaged. Fifteen other homes on North Island were splintered and wrecked by the waves. The town pier was significantly damaged and the popular visitor's lot at Chatham Light was undermined. Throughout other parts of the Cape, beach stairs, piers, and boats were swept away to be deposited back on the shore in a new spot. All this was just two months after Hurricane Bob had brought even stronger winds to the area for a one-two punch of destructive weather.

On Martha's Vineyard roads were turned into rivers and the flooding was some of the worst ever witnessed by longtime islanders. The Tisbury

Wharf in Vineyard Haven was destroyed by the wave action and ferries were shut down for days.

Mark Alan Lovewell, reporter at the *Vineyard Gazette*, described the storm as one with the power of a winter northeaster that came in "raging across barrier beaches and sandy island perimeter with flood tides not seen since the double hurricanes of 1954." The hurricanes he was speaking of were Carol and Edna, which also earn spots on the list of New England's worst storms.

No strangers to big flooding, there are many measuring posts revealing high marks of the past in places like Martha's Vineyard. One of those is at the Edgartown Yacht Club, where the Perfect Storm's tide topped both 1954 hurricanes and came short of only the Great Atlantic Hurricane of 1944 and the Hurricane of '38, which bought unbelievable storm surge into the south coasts of New England.

Whatever defenses that had been weakened by Hurricane Bob back in August were brought down by the incessant waves and wind. Many beaches lost twenty to fifty feet of sand and frontage, making it one of the largest single erosion events from any coastal storm.

Up the shoreline to Maine, an iconic location was battered and bruised. Located on scenic Walker's Point in Kennebunkport was the summer home of President George H. W. Bush and family. The fetching compound on trademark craggy rocks of Maine's coast was inundated by water and debris. The property was so badly flooded that all the land was covered and only the house stuck up above the water like a buoy asking for help. As the water receded, the driveway looked like a bomb went off with rocks and other assorted debris strewn all over the road. Upon hearing the news, the President had a humble reply. "It was rather devastating to our family, but when I compare that to the fortunes of others, why, we've got a lot to be grateful for."

The Bush family wasn't alone as more than one hundred homes were damaged on the southern Maine coast. Governor John R. McKernan declared a state of emergency for the coastal zones of Cumberland and York Counties. Bad as it was, the flooding could have been even worse. The Perfect Storm struck about a week after the highest tides of the month. If it had arrived about five days earlier, water levels would have been as much as a foot higher.

While oceanside dwellers were in the throes of one of their most memorable storms, the impact scaled back to near zero once inland. It was rainy and breezy, but most anyone who had no interest or ties to the ocean would not remember the event as more than just a wet Halloween. It was a coastal storm in the purest sense.

The strangest and most unique part of the Perfect Storm's story occurred at the very end of its evolution. From a cold front to an extra-tropical nor'easter that absorbed a hurricane, it gave birth to yet another hurricane. If you can follow this progression, congrats, you are now an honorary meteorologist.

On Halloween, the core of the waning nor'easter had drifted just east of Norfolk, Virginia. The sprawling system had made its way down to the warm waters of the Gulf Stream now and began taking on subtropical characteristics. Even at the end of October, it is plenty warm enough over the Gulf Stream (approximately 80°F) to support tropical cyclones. And so it was. A full-fledged hurricane, albeit small in size, all of a sudden popped into existence on November 1.

NOAA Hurricane Hunters made their way out to this oddity and found it to have a central pressure of 980 millibars with flight-level winds of ninety-nine miles per hour. That was good enough to be classified a Category 1 hurricane with seventy-five-miles-per-hour winds at the surface. It was the eighth hurricane of the 1991 Atlantic season.

If it became a true hurricane, why does it have no name? If you look into the archives of the National Hurricane Center (NHC), there is a listing and track for an "unnamed hurricane" but not Hurricane Henri, which was the next name on the list that year.

It was all about communication. As the tropical transition was ongoing, all attention and media coverage was on coastal damage and the search for the *Andrea Gail*. The prevailing thought among forecasters was that adding a new hurricane into the mix would have been very confusing for the public and emergency managers. Especially since the new storm was not expected to worsen any impacts for those already trying to clean up.

So a decision was made between NOAA's National Meteorological Center, several National Weather Service Forecast Offices, and the US Navy to leave it unnamed. It was tracked and people in the Canadian Maritimes were still warned that a storm was coming, just without a trademark tropical name. The center went on to head north and make landfall in Nova Scotia as a weakening tropical storm on November 2. Thus ended the complex and strange life of the Perfect Storm.

Sources

"The Children's Blizzard, January 12, 1888." arcgis.com. National Weather Service—Aberdeen, SD. https://www.arcgis.com/apps/MapSeries/index. html?appid=3b68adee4e9545b7abdd7355ab7fe367.

Liepa, Bob. "80 Years Ago, No One Saw the Hurricane of 1938 Coming." suffolktimes.timesreview.com. *The Suffolk Times*, January 3, 2020. https:// suffolktimes.timesreview.com/2018/09/north-fork-history-project-80 -years-ago-no-one-saw-hurricane-1938-coming.

"10th Anniversary of the Devastating 1998 Ice Storm in the Northeast." IceStorm1998.pdf, January 5, 2008. https://www.weather.gov/media/btv /events/IceStorm1998.pdf.

"1815—The Great September Gale." Hurricanes: Science and Society. University of Rhode Island, 2020. http://hurricanescience.org/history/ storms/pre1900s/SeptemberGale.

"1816: The Year without a Summer." newenglandhistoricalsociety.com. New England Historical Society, December 6, 2020. https://www.newengland historicalsociety.com/1816-year-without-a-summer.

"The 1936 Flood That Engulfed New England." newenglandhistoricalsociety. com. New England Historical Society, March 11, 2021. https://www.new englandhistoricalsociety.com/great-new-england-flood-1936.

"The 1947 Fire." kennebunkportme.gov. Town of Kennebunkport, ME. https:// www.kennebunkportme.gov/fire-department/pages/1947-fire.

"The 1947 Fire." waterboro-me.net. The Waterborough Historical Society, October 1978. www.waterboro-me.net/document_center/history /fire_1947.pdf.

"The 1947 Fires." maine.gov. Maine Forest Service. https://www.maine.gov /dacf/mfs/forest_protection/1947_fire.html.

"The 1998 Ice Storm: 10-Year Retrospective." rms.com. Risk Management Solutions, Inc., 2008. https://forms2.rms.com/rs/729-DJX-565/images /wtr_1998_ice_storm_10_retrospective.pdf.

"60th Anniversary of Hurricane Edna." noaahrd.wordpress.com. Hurricane Research Division/NOAA/OAR/Atlantic Oceanographic and Meteorological Laboratory, September 10, 2014. https://noaahrd .wordpress.com/2014/09/10/60th-anniversary-of-hurricane-edna.

"About the NFFPC." nffpc.org. Northeastern Forest Fire Protection Commission. https://www.nffpc.org/en/information/about.

Ahles, Dick. "Memories of a Flood, 50 Years Later." nytimes.com. *The New York Times*, August 14, 2005. https://www.nytimes.com/2005/08/14 /nyregion/memories-of-a-flood-50-years-later.html.

"Al Southwick: The Hottest of Summers—1911." telegram.com. telegram.com, July 26, 2017. https://www.telegram.com/opinion/20170727/al-southwick -hottest-of-summers---1911.

Andersen, Travis. "'Three Days of Hell'—Scars of Blizzard of '78 Linger, 40 Years Later—The Boston Globe." BostonGlobe.com. *The Boston Globe*, February 2, 2018. https://www.bostonglobe.com/2018/02/02/years-later -they-shudder-thought-blizzard/fYA7dYGuGVQIgjly8R2GCP/story .html.

Andrews, Evan. "Remembering New England's 'Dark Day.'" History.com. A&E Television Networks, May 19, 2015. https://www.history.com/news /remembering-new-englands-dark-day.

"As Jet Is about to Land on Tornado, It's Not Noisy When All Are Praying." news.google.com. *Spokane Daily Chronicle*, October 5, 1979. https://news .google.com/newspapers?id=FqQSAAAAIBAJ&sjid=LvkDAAAAIBAJ &pg=1882%2C1564816.

The Associated Press. "Bush to Assess Damage to Kennebunkport Home Battered by Sea." google.com. *The Pittsburgh Press*, October 31, 1991. https://news.google.com/newspapers?id=Md8cAAAAIBAJ&sjid=GGQE AAAAIBAJ&dq=storm+new+england&pg=3692%2C8275025.

The Associated Press. "Hartford, CT Flood, Mar 1936—Swirling Waters." gendisasters.com. GenDisasters.com, March 20, 1936. http://www.gen disasters.com/connecticut/7022hartford-ct-flood-mar-1936-swirling -waters.

The Associated Press. "Ice Storm Cripples Parts of Northeast (Published 2008)." nytimes.com. *The New York Times*, December 13, 2008. https:// www.nytimes.com/2008/12/12/us/12cnd-storm.html.

The Associated Press. "Wind and Water Take Toll along Connectiut Shore." google.com. *Record-Journal*, November 1, 1991. https://news.google .com/newspapers?id=ihxIAAAAIBAJ&sjid=rgANAAAAIBAJ&pg=1620 %2C180883.

The Associated Press. "New England Ready for Date with Edna." google.com. *The New London Day*, September 11, 1954. https://news.google.com/newsp

apers?id=Ue8gAAAAIBAJ&sjid=9nIFAAAAIBAJ&dq=hurricane+edna&
pg=2114%2C1735252.

Austin, Nick. "How Deadly 1888 Blizzard Transformed US Public
Transportation." FreightWaves, March 12, 2020. https://www.freightwaves
.com/news/how-deadly-1888-blizzard-transformed-us-public-trans
portation.

Avila, Lixion A., and John Cangialosi. "Tropical Cyclone Report Hurricane
Irene." nhc.noaa.gov. National Hurricane Center, April 11, 2013. https://
www.nhc.noaa.gov/data/tcr/AL092011_Irene.pdf.

Bacon, Kezia. "120 Years Ago: The Portland Gale." nsrwa.org. North and South
Rivers Watershed Association, May 1, 2019. https://www.nsrwa.org/120
-years-ago-portland-gale.

Barron, James. "One Casualty of Northeaster: 'Trick or Treat!'" The New York
Times. *The New York Times*, October 31, 2011. https://www.nytimes.com
/2011/11/01/nyregion/1-million-still-in-dark-after-destructive-weekend
-storm.html.

"Bartonsville Covered Bridge." bartonsvillecoveredbridge.com. Bartonsville
Covered Bridge, 2021. https://www.bartonsvillecoveredbridge.com/history.

Beitler, Stu. "Bar Harbor, ME Forest and Town Fires, Oct 1947." gendisasters
.com. http://www.gendisasters.com/maine/9474/bar-harbor-me-forest
-town-fires-oct-1947.

Beitler, Stu. "MA, NY, NC Hurricane EDNA, Sept 1954." gendisasters.com.
GenDisasters, September 11, 1954. https://www.gendisasters.com/new
-york/6392/ma-ny-nc-hurricane-edna-sept-1954?page=0%2C2.

Blake, Eric S., Edward N. Rappaport, Jerry D. Jarrell, and Christopher W.
Landsea. "The Deadliest, Costliest, and Most Intense United States
Tropical Cyclones from 1851 to 2004." www.aoml.noaa.gov. Tropical
Prediction Center, National Hurricane Center, Miami, Florida, August
2005. https://www.aoml.noaa.gov/general/lib/lib1/nhclib/nwstechmemos
/TechMemoTPC%234.pdf.

"The Blizzard of '78—U.S. National Archives—Google Arts & Culture."
Google. US National Archives. https://artsandculture.google.com/exhibit
/the-blizzard-of-78-u-s-national-archives/QRSaaAEN?hl=en.

"The Blizzard of '78—Forty Years Later." CapeCod.com, February 2, 2018.
https://www.capecod.com/lifestyle/the-blizzard-of-78-forty-years-later.

"The Body of Boy Found as Snow Melts." The Hour. Google. https://news
.google.com/newspapers?nid=1916&dat=19780301&id=bu8pAAAAIBAJ
&sjid=Pm4FAAAAIBAJ&pg=1075%2C228300.

Bogart, Dean B. "Floods of August–October 1955 New England to North Carolina." Geological Survey Water-supply Paper. United States Geological Survey. https://books.google.com/books?id=wNAPAAAAIAA J&pg=PA1&dq=hurricane%2Bconnie%2B1955&hl=en#v=onepage&q=hur ricane%20connie%201955&f=false.

Boston, NWS. "The Portland Gale—November 26th–27th, 1898." Twitter. National Weather Service, November 26, 2017. https://twitter.com/nws boston/status/934783892959256576.

Bothwell, Dick. "Something's to Be Done about Weather." google.com /newspapers. *St. Petersburg Times*, September 26, 1960. https://news .google.com/newspapers?id=jD5SAAAAIBAJ&sjid=DXkDAAAAIBAJ &pg=4035%2C2831370&dq=hurricane%2Bdiane%2B1955&hl=en.

Bourcier, Paul G., and Robert P. Emlen. "Rhode Island History." rihs.org. Rhode Island Historical Society, May 1990. www.rihs.org/assetts/files /publications/1990_May.pdf.

"Brattleboro Historical Society: This Week, 83 Years Ago . . ." reformer.com. *Brattleboro Reformer*, March 22, 2019. https://www.reformer.com /community-news/brattleboro-historical-society-this-week-83-years-ago /article_1f430c51-5503-5380-b24e-c8716ac78e8f.html.

"A Brief History of PCC—Poquonock Community Church." pccwindsor.com. Poquonock Community Church. https://www.pccwindsor.com/history .html.

Brown, Keith L. "Joseph Smith Family and the Year without a Summer." historyofmormonism.com. History of Momonism, March 12, 2015. https://historyofmormonism.com/2015/03/12/joseph-smith-family-and -the-year-without-a-summer.

Brown, Tom. "Marker Chronicles City's Flood History." montpelierbridge .org. the bridge, November 8, 2019. https://montpelierbridge.org/2019/11 /marker-chronicles-citys-flood-history.

Burnham, Emily. "70 Years Ago This Month, the Great Fires of 1947 Raged across Maine." bangordailynews.com. *Bangor Daily News*, October 6, 2017. https://bangordailynews.com/2017/10/05/arts-culture/70-years-ago -this-month-the-great-fires-of-1947-raged-across-maine.

Burt, Christopher C. "The Blizzard of 1888: America's Greatest Snow Disaster." Weather Underground, March 12, 2020. https://www.wunderground.com /cat6/the-blizzard-of-1888-americas-greatest-snow-disaster.

Bushnell, Mark. "Then Again: Remembering the Terror and Losses of 1927 Flood." vtdigger.org. VTDigger, September 4, 2016. https://vtdigger.org

/2016/09/04/then-again-remembering-the-terror-and-losses-of-1927
-flood.

"Canadian Tropical Cyclone Season Summary for 1954." ec.gc.ca, January 15,
2015. https://www.ec.gc.ca/ouragans-hurricanes/default.asp?lang=en
&n=7AC77512-1.

Cavanaugh, Ray. "Providence Besieged by Great Gale in 1815." providence
journal.com. *Providence Journal*, September 17, 2015. https://www
.providencejournal.com/article/20150920/entertainmentlife/150929983.

CBS Boston. "5 Things You Didn't Know about WBZ NewsRadio 1030's
History." boston.cbslocal.com. CBS Boston, June 27, 2013. https://boston
.cbslocal.com/top-lists/5-things-you-didnt-know-about-wbz-newsradio
-1030s-history.

"Chilly Temperatures during the Maunder Minimum." earthobservatory.nasa
.gov. NASA. https://earthobservatory.nasa.gov/images/7122/chilly-tem
peratures-during-the-maunder-minimum.

Coffin, Joshua. "A Sketch of the History of Newbury, Newburyport, and West
Newbury, from 1635 to 1845. (1845 Edition)." Open Library. S. G. Drake,
January 1, 1970. https://openlibrary.org/books/OL6905506M/A_sketch
_of_the_history_of_Newbury_Newburyport_and_West_Newbury_from
_1635_to_1845.

Cotterly, Wayne. "Hurricane Edna-(1954)." pivot.net, 2002. https://web.archive
.org/web/20061220040916/http://www.pivot.net/~cotterly/edna.htm.

Curley, Ryan. "Blizzard of 1978 at Coast Guard Beach, Cape Cod." TheFuries.
https://www.thefuriesonline.com/blizzard-1978-cape-cod
-coast-guard-beach.

"Dam History." westhillpond.org. West Hill Pond Association. https://westhill
pond.org/about-us/dam-history.

"Damage Set at 7 Million by Governor." Google News Archive Search. *The
Lewiston Daily Sun*, September 13, 1954. https://news.google.com/news
papers?id=okMpAAAAIBAJ&sjid=zWYFAAAAIBAJ&pg=6104%2C502
8453&dq=hurricane%2Bedna&hl=en.

Damon, Laura. "Hurricane of '38 Left Much of New England Shattered."
providencejournal.com. *The Providence Journal*, September 20, 2018.
https://www.providencejournal.com/news/20180920/hurricane-of-38
-left-much-of-new-england-shattered.

DeCosta-Klipa, Nik. "When the Blizzard of '78 Shut down Boston, the Garden
Stayed Open." Boston.com. *The Boston Globe*, January 30, 2018. https://

www.boston.com/news/history/2018/01/30/blizzard-of-78-beanpot
-boston-garden.

DeGaetano, Arthur T. "Climatic Perspective and Impacts of the 1998
Northern New York and New England Ice Storm." JSTOR. American
Meteorological Society, February 2000. https://www.jstor.org.stable
/26215101.

D'Entremont, Jeremy. "Thacher Island Twin Lights History." New England
Lighthouses: A Virtual Guide. http://www.newenglandlighthouses.net
/thacher-island-twin-lights-history.html.

Dinsmore, Brad. "Great Snow of 1717." Windham NH History, February 18,
2015. https://windhamnhhistory.org/category/great-snow-of-1717.

"Documentation of Atlantic Tropical Cyclones Changes in HURDAT." www
.aoml.noaa.gov. NOAA's Atlantic Oceanographic and Meteorological
Laboratory. https://www.aoml.noaa.gov/hrd/hurdat/metadata_dec12.html.

Dolce, Chris, and Jon Erdman. "The Nation's Worst Ice Storms: The Weather
Channel—Articles from The Weather Channel." The Weather Channel,
January 11, 2017. https://weather.com/storms/winter/news/top-10-worst
-ice-storms-20131205.

Donnelly, Jeffrey P. "700 Yr. Sedimentary Record of Intense Hurricane
Landfalls in Southern New England." www.geo.brown.edu. Brown
University, June 2001. www.geo.brown.edu/georesearch/esh/QE/
Publications/GSAB2001
/JDonnelly/Succotash/Succotach.pdf.

Drake, Samuel Adams. "Thacher's Island and Anthony Thacher's Shipwreck."
The Shipwreck of Anthony Thacher—1635. Roberts Brothers. https://
troutwind.tripod.com/shipwre.html.

Dunbar, Brian. "Volcanic Blast Location Influences Climate Reaction." nasa
.gov. NASA. https://www.nasa.gov/centers/goddard/news/topstory/2005
/volcano_climate.html.

Dunlop, Tom. "In the Eye of the Hurricane: Edna on the Heels of Carol."
vineyardgazette.com. *The Vineyard Gazette*—Martha's Vineyard News,
September 11, 2014. https://vineyardgazette.com/news/2014/09/11/eye
-hurricane-edna-heels-carol.

Dupigny-Giroux, Lesley-Ann. "Impacts and Consequences of the Ice Storm
of 1998 for the North American North-East." rmets.onlinelibrary. Royal
Meteorological Society, April 30, 2012. https://rmets.onlinelibrary.wiley
.com/doi/pdf/10.1002/j.1477-8696.2000.tb04012.x.

Dwyer, Dialynn. "'The Year That Maine Burned': 70 Years Ago Ravaging Fires Left Thousands of Mainers Homeless." Boston.com. *The Boston Globe*, October 29, 2017. https://www.boston.com/news/history/2017/10/29/the -year-that-maine-burned-70-years-ago-ravaging-fires-left-thousands-of -mainers-homeless.

Eckholm, Erik. "Covered Bridges, Beloved Remnants of Another Era, Were Casualties, Too." nytimes.com. *The New York Times*, September 1, 2011. https://www.nytimes.com/2011/09/01/us/01bridges.html.

Editors of *Encyclopaedia Britannica*. "Mount Tambora." britannica.com. Encyclopædia Britannica, inc. https://www.britannica.com/place /Mount-Tambora.

Enman, Steve. "Berlin Residents Remember Arena Collapse." conwaydailysun .com. *The Conway Daily Sun*, February 27, 2019. https://www.conway dailysun.com/news/local/berlin-residents-remember-arena-collapse /article_9cb934a2-3a03-11e9-a9ce-b752326f55af.html.

Evans, Jean B. "Blizzard of 1898, the Portland Gale." Mystic and Stonington—A Storied Past and Present! Connecticut Tales from my Ancestors, January 7, 2012. https://mystoningtonancestors.wordpress .com/2012/01/07/blizzard-of-1898-the-portland-gale.

Evans, Robert. "Blast from the Past." Smithsonian.com. *Smithsonian Magazine*/Smithsonian Institution, July 1, 2002. https://www.smithsonian mag.com/history/blast-from-the-past-65102374.

"Exeter: Hurricanes Carol, Edna Strike in 1954." WMUR.com. WMUR, September 24, 2019. https://www.wmur.com/article/exeter-hurricanes -carol-edna-strike-in-1954/5165383.

Fankhauser, Sean. "A Look Back at Hurricane Carol (1954)." bluehill.org. Blue Hill Observatory & Science Center, August 30, 2013. http://bluehill.org /observatory/2013/08/a-look-back-at-hurricane-carol-1954.

Finch, Robert. "Remembering the 'Perfect Storm' 25 Years Later." capeand islands.org. CAI, October 29, 2016. https://www.capeandislands.org/in -this-place/2016-10-25/remembering-the-perfect-storm-25-years-later #stream/0.

"Fire of 1947." mdihistory.org. Mount Desert Island Historical Society. http:// mdihistory.org/exhibits/htdocs/fire.

"Fire of 1947." nps.gov. National Parks Service: U.S. Department of the Interior. https://www.nps.gov/acad/learn/historyculture/fireof1947.htm.

Fisher, Eric. "Eye on Weather: Lessons Learned from Hurricane of '38." boston.cbslocal.com. CBS Boston, May 21, 2016. https://boston.cbslocal

.com/2016/05/21/eye-on-weather-hurricane-of-1938-rhode-island-new
-england-long-island-express.

"Flood Devastates Springfield." massmoments.org. Mass Moments, March 21,
1936. https://www.massmoments.org/moment-details/flood-devastates
-springfield.html.

"The Flood of '27, 1927." vermonthistory.org. Vermont Historical Society.
https://vermonthistory.org/flood-of-271927?web=1&wdLOR=cE658BF51
-1FC3-3940-B65B-FFD05D05CFDA.

"Flooding in Vermont." weather.gov. National Oceanic and Atmospheric
Administration's National Weather Service, April 19, 2018. https://www
.weather.gov/safety/flood-states-vt.

Freedman, Andrew. "Historic October Northeast Storm: Epic. Incredible.
Downright Ridiculous." *The Washington Post*. WP Company, October 31,
2011. https://www.washingtonpost.com/blogs/capital-weather-gang/post
/historic-october-northeast-storm-epic-incredible-downright-ridiculous
/2011/10/31/gIQApy7LZM_blog.html.

Freedman, Andrew. "Hurricane Irene Ranked Most Costly Category 1 Storm."
climatecentral.org. Climate Central, May 10, 2012. https://www.climate
central.org/news/hurricane-irene-ranked-most-costly-category-1-storm.

Gottlieb, Jane. "In Vermont, a Town That Would Not Let Its Diner Go."
nytimes.com. *The New York Times*, December 14, 2013. https://www
.nytimes.com/2013/12/14/us/in-vermont-a-town-that-would-not-let-its
-diner-go.html.

"The Great 1938 Hurricane, a Once-in-a-Lifetime Storm." newengland
historicalsociety.com. New England Historical Society, March 3, 2021.
https://www.newenglandhistoricalsociety.com/great-1938-hurricane.

"The Great New England Hurricane of 1938—History." weather.gov. NOAA's
National Weather Service, September 18, 2020. https://www.weather.gov
/okx/1938HurricaneHistory.

"The Great September Gale of 1815." newenglandhistoricalsociety.com. New
England Historical Society, September 22, 2020. https://www.newengland
historicalsociety.com/the-great-september-gale-of-1815.

"The Great Vermont Flood of 1927, November 3–4." weather.gov. National
Oceanic and Atmospheric Administration/National Weather Service.
https://www.weather.gov/media/btv/events/1927Flood.pdf.

Gregersen, Erik. "Dalton Minimum: Solar Phenomenon [1790–1830]."
britannica.com. Encyclopædia Britannica, Inc., September 23, 2011.
https://www.britannica.com/science/Dalton-minimum.

"Groton Impacts." weather.gov. https://www.weather.gov/media/okx/coastal flood/Groton%20impacts.pdf.

Grover, Nathan C. "The Floods of March 1936." pubs.usgs.gov. United States Department of the Interior/Federal Emergency Administration of Public Works, 1937. https://pubs.usgs.gov/wsp/1779m/report.pdf.

"The Halloween Nor'Easter of 1991: East Coast of the United States . . . Maine to Florida and Puerto Rico." weather.gov. National Oceanic and Atmospheric Administration/National Weather Service. https://www.weather.gov/media/publications/assessments/Halloween%20Nor'easter%20of%201991.pdf.

Hanrahan, Ryan. "A New Look to Long Island Sound Beaches." Ryan Hanrahan.com, September 3, 2011. https://www.ryanhanrahan.com/tag/hurricane-irene/page/2.

Hanrahan, Ryan. "Flood of August 1955." RyanHanrahan.com. http://www.ryanhanrahan.com/flood-of-august-1955.

Hanrahan, Ryan. "Here Are the Original Coop Forms from Middletown and Wallingford, CT from the Blizzard of 1888. 42" in Middletown and 50" in Wallingford. What's Crazy Is the 5"–6" of Liquid Equivalent Too! Pic. twitter.com/tuU173n35l." Twitter. Twitter, March 12, 2019. https://twitter.com/ryanhanrahan/status/1105491430326308864.

Harris, Gordon. "The Dark Day, May 19, 1780." historicipswich.org. Historic Ipswich, September 17, 2020. https://historicipswich.org/2020/05/15/the-dark-day-1780.

Henderson-Shifflett, Jeannine. "Blizzard of 1888 Devastates State: Connecticut History: A CTHumanities Project." Connecticut History/CTHumanities Project, December 17, 2018. https://connecticuthistory.org/blizzard-of-1888-devastates-state.

Herbert, Paul N. "May 19, 1780—New England's Dark Day." massar.org. The Massachusetts Society Sons of the American Revolution, March 9, 2013. https://www.massar.org/2013/03/09/may-19-1780-new-englands-dark-day.

Hickey, Walter V. "The Final Voyage of the *Portland*." archives.gov. National Archives and Records Administration, 2006. https://www.archives.gov/publications/prologue/2006/winter/portland.html.

Hidlay, William C. "Maine Hit Hard by Storm." google.com. *Bangor Daily News*, November 1, 1991. https://news.google.com/newspapers?id=DqhJA AAAIBAJ&sjid=KQ4NAAAAIBAJ&pg=3345%2C8099&dq=storm%2B new%2Bengland&hl=en.

"Historic Flood March 1936." weather.gov. NOAA's National Weather Service, September 21, 2015. https://www.weather.gov/nerfc/hf_march_1936.

"Historical Timeline." bluehill.org. Blue Hill Observatory & Science Center. https://bluehill.org/observatory/about-us/historical-timeline.

"History of the American Clock Business for the Past Sixty Years." Google Books. Google. https://books.google.com/books?id=MzoZAAAAYAAJ &printsec=frontcover&dq=chauncey%2Bjerome%2Bhistory%2Bof%2B clock&hl=en&sa=X&ei=VeaRU-bmFs-Aqgbkl4DgAg#v=onepage&q =chauncey%20jerome%20history%20of%20clock&f=false.

"How Volcanoes Influence Climate." scied.ucar.edu. University Corporation for Atmospheric Research. https://scied.ucar.edu/learning-zone/how-climate -works/how-volcanoes-influence-climate.

https://www.wpc.ncep.noaa.gov. Weather Prediction Center/National Centers for Environmental Prediction/National Oceanic and Atmospheric Administration, October 30, 2011. /winter_storm_summaries/event _reviews/2011/Autumn_Mid-Atlantic_to_Northeast_US_Winter_Storm _Oct2011.pdf.

"Hurricane and Tropical Cyclones." wunderground.com. Weather Under-ground. https://www.wunderground.com/hurricane/atlantic/1991 /Hurricane-Grace.

"Hurricane Carol Causes Heavy Loss in Seacoast Area." hampton.lib.nh.us. The Hampton Union & Rockingham County Gazette/Lane Memorial Library. http://www.hampton.lib.nh.us/hampton/history/storms/1954hurricane CarolHU19540902.htm.

"Hurricane Carol, So Deadly Her Name Was Retired." newenglandhistorical society.com. New England Historical Society, September 20, 2020. https://www.newenglandhistoricalsociety.com/hurricane-carol-deadly -name-retired.

"Hurricane Connie, August 12, 1955." weather.gov. NOAA's National Weather Service, June 13, 2015. https://www.weather.gov/mhx /HurricaneConnie1955.

"Hurricane Diane, 1st $1 Billion Hurricane, Wallops New England in 1955." newenglandhistoricalsociety.com. New England Historical Society, July 11, 2020. https://www.newenglandhistoricalsociety.com/hurricane-diane-1st -1-billion-hurricane-wallops-new-england-1955.

"Hurricanes Connie & Diane Deliver Double Hit—Who Knew?" connecticuthistory.org. Connecticut History/CTHumanities Project,

August 9, 2020. https://connecticuthistory.org/hurricanes-connie-diane
-deliver-double-hit-who-knew.

"Ice Storm Genealogy Project." geni_family_tree. https://www.geni.com
/projects/Ice-Storm/40039.

"Irene Graphics Archive." www.nhc.noaa.gov. National Hurricane Center/
National Oceanic and Atmospheric Administration. https://www.nhc
.noaa.gov/archive/2011/graphics/al09/loop_5W.shtml.

"Irene's Wrath Is Felt in the Valley." valleyreporter.com. *The Valley Reporter*,
December 29, 2011. https://www.valleyreporter.com/index.php/en/83
-news/285-irenes-wrath-is-felt-in-the-valley.

Jarvinen, Brian R. "Storm Tides in Twelve Tropical Cyclones (Including Four
Intense New England Hurricanes)." www.aoml.noaa.gov. NOAA/Tropical
Prediction Center/National Hurricane Center, October 1, 2006. https://
www.aoml.noaa.gov/hrd/Landsea/12Tides.pdf.

Jones, Brian C. "A Defining Event in R.I. History." projo.com. *The Providence
Journal*—Weather. https://web.archive.org/web/201106290453
48/http:/www.projo.com/cgi-bin/include.pl/specials/blizzard/story1.htm.

Jones, Kathleen F. "The December 2008 Ice Storm in New Hampshire." puc.
nh.gov. Cold Regions Research and Engineering Laboratory, n.d. https://
www.puc.nh.gov/2008IceStorm/Final%20Reports/2009-10-30%20
Final%20NEI%20Report%20With%20Utility%20Comments/Appendix
%20D%20-%20CRREL%20Report.pdf.

Judge, Josh. "200 Years Ago, Summer Didn't Come to New Hampshire."
WMUR.com. WMUR, February 26, 2021. https://www.wmur.com
/article/200-years-ago-summer-didn-t-come-to-new-hampshire/5210654.

Kessler, Edwin. "Eye Region of Hurricane Edna, 1954." AMETSOC.org.
American Meteorological Society, June 1, 1958. https://journals.ametsoc
.org/view/journals/atsc/15/3/1520-0469_1958_015_0264_erohe_2_0
_co_2.xml.

Kinnison, H. B. "The New England Flood of November 1927." usgs.gov.
United States Geological Survey. https://pubs.usgs.gov/wsp/0636c/report
.pdf.

Kocin, Paul J. "An Analysis of the 'Blizzard of '88.'" AMETSOC. American
Meteorological Society, November 1, 1983. https://journals.ametsoc.org
/view/journals/bams/64/11/1520-0477_1983_064_1258_aaoto_2_0_co_2
.xml.

Landsea, Chris, Mike Dickinson, and Donna Strahan. "Reanalysis of Ten
U.S. Landfalling Hurricanes." noaa.gov. National Oceanic and

Atmospheric Administration. https://www.aoml.noaa.gov/hrd/hurdat
/10_US_hurricanes.pdf.

Livingston, Ian. "Violent F4/EF-4 and F5/EF-5 Tornadoes in the United States Since 1950." ustornadoes.com, February 17, 2016. https://www .ustornadoes.com/2012/04/10violent-f4ef-4-and-f5ef-5-tornadoes-in-the -united-states-since-1950.

Long, Stephen. "One for the Ages: The Hurricane of 1938 Battered New England's Woods 75 Years Ago: Articles: Features." Center for Northern Woodlands Education. Northern Woodlands. https://northernwoodlands .org/articles/article/hurricane-1938.

Lovewell, Mark Alan. "Tidal Surges, Winds Turn Vineyard into New Disaster Zone." vineyardgazette.com. *The Vineyard Gazette*—Martha's Vineyard News, November 1, 1991. https://vineyardgazette.com/news/1991/11/01 /tidal-surges-winds-turn-vineyard-new-disaster-zone.

Lowney, Mary Ellen. "40 Years Ago: Tornado Killed 3, Destroyed Bradley Air Museum and Ravaged Region." masslive.com. MassLive, October 13, 2019. https://www.masslive.com/living/2019/10/40-years-ago-tornado-killed-3 -destroyed-bradley-air-museum-and-ravaged-region.html.

Lubchenco, Jane, and Laura Furgione. "Hurricane Irene, August 21–30, 2011 U.S." weather.gov. U.S. Department of Commerce/National Oceanic and Atmospheric Administration, September 2012. https://www.weather.gov /media/publications/assessments/Irene2012.pdf.

Mahony, Edmund H. "Extreme Weather of 2011: Freak October Snowstorm." courant.com. *Hartford Courant*, December 13, 2018. https://www.courant .com/news/connecticut/hc-halloween-year-1228-20111223-story.html.

Margasak, Larry, and Kathy Morisse. "The Blizzard of 1888." National Museum of American History, January 4, 2018. https://americanhistory.si.edu/blog /blizzard-1888.

Martin, Vivian B. "Body of Victim Discovered in Rubble." Newspapers.com. *Hartford Courant*, October 5, 1979. https://www.newspapers.com/image /?clipping_id=24228261&fcfToken=eyJhbGciOiJIUzI1NiIsInR5cCI6Ikp XVCJ9.eyJmcmVlLXZpZXctcaWQiOjM2ODM3NT M5NCwiaWF0I joxNTkxNzUxMzY4LCJleHAiOjE1OTE4Mzc3Njh9.enklp1AP7V5fAV F6EPp7oqeJbN81uGj9RApJr7QO4PM.

Marvel, William. "The Day Brownfield Burned." conwaydailysun.com. *The Conway Daily Sun*, October 17, 2017. https://www.conwaydailysun.com /news/local/the-day-brownfield-burned/article_c17d8b18-aade-11e7 -aa99-bf33a83391cd.html.

"Mary Shelley Biography." Biography.com. A&E Networks Television, February 28, 2020. https://www.biography.com/writer/mary-shelley.

Mathews, Amanda A. "'Covered with Egyptian Darkness': New England's Dark Day of 1780." masshist.org. The Beehive/Massachusetts Historical Society, May 13, 2015. https://www.masshist.org/beehiveblog/2015/05/covered-with-egyptian-darkness-new-englands-dark-day-of-1780.

"Mattapoisett and Old Rochester, Massachusetts—Buzzards Bay." buzzardsbay.org. https://buzzardsbay.org/download/1847-rochester-history-1815-gale.pdf.

Maye, Brian. "A Volcanic Eruption with Global Repercussions—An Irishman's Diary on 1816, the Year without a Summer." irishtimes.com. *The Irish Times*, August 19, 2016. https://www.irishtimes.com/opinion/a-volcanic-eruption-with-global-repercussions-an-irishman-s-diary-on-1816-the-year-without-a-summer-1.2760797.

McBurney, Christian. "The Great Gale of 1815 Slams into Newport, Providence and Narragansett." smallstatebighistory.com. The Online Review of Rhode Island History. http://smallstatebighistory.com/the-great-gale-of-1815-slams-into-newport-providence-and-narragansett.

McCown, Sam. "The Perfect Storm: October 1991." ncdc.noaa.gov. National Oceanic and Atmospheric Administration/National Climatic Data Center, January 9, 2017. https://web.archive.org/web/20170109135251/https://www.ncdc.noaa.gov/oa/satellite/satelliteseye/cyclones/pfctstorm91/pfctstorm.html.

McCown, Sam. NCDC: Satellite Events Art Gallery: Hurricanes. https://web.archive.org/web/20161212000841/https://www.ncdc.noaa.gov/oa/satellite/satelliteseye/hurricanes/unnamed91/unnamed91.html.

McDonough, Kerry. "A Tale of Three Steeples." oldnorth.com. The Old North Church, August 31, 2015. https://www.oldnorth.com/blog/a-tale-of-three-steeples.

Mcgeehan, Patrick. "Days after the Storm, Many Are Still in the Dark." nytimes.com. *The New York Times*, September 1, 2011. https://www.nytimes.com/2011/09/01/nyregion/days-after-the-storm-many-are-left-in-the-dark.html?scp=6&sq=irene+connecticut&st=cse.

McGinnes, Meagan. "25 Years Ago, the Crew of the *Andrea Gail* Was Lost in the 'Perfect Storm.'" boston.com. https://www.boston.com/news/history/2016/10/29/25-years-ago-the-crew-of-the-andrea-gail-were-lost-in-the-perfect-storm.

McLain, Guy. "The Flood of 1936." springfieldmuseums.org. Springfield Museums, March 30, 2016. https://springfieldmuseums.org/blog/flood -of-1936.

McMurry, Erin R., Michael C. Stambaugh, Richard P. Guyette, and Daniel C. Dey. "Fire Scars Reveal Source of New England's 1780 Dark Day." publish .csiro.au. Csiro Publishing, July 3, 2007. https://www.publish.csiro.au/wf /WF05095.

Miller-Weeks, Margaret, Chris Eagar, and Christina M. Petersen. "The Northeastern Ice Storm 1998, a Forest Damage Assessment for New York, Vermont, New Hampshire, and Maine." fs.fed.us. USDA Forest Service, December 1999. https://www.fs.fed.us/nrs/pubs/jrnl/1999/ne_1999_miller -weeks_001.pdf.

Mitchell, Jennifer. "Maine May Have 'Year without a Summer' to Thank for Its Statehood." mainepublic.org. Maine Public Radio, July 27, 2016. https:// www.mainepublic.org/post/maine-may-have-year-without-summer -thank-its-statehood.

Moore, Tom. "The Northeast 'Dark Day' in May of 1780 Spread Fear and Bewilderment." weatherconcierge.com. Weather Concierge, May 16, 2019. https://www.weatherconcierge.com/the-northeast-dark-day-in-may-of -1780-spread-fear-and-bewilderment.

Moore, Tom. "'The Great Snow' in 1717 Created Great Hardships in the Colonies." Weather Concierge, March 6, 2020. https://www.weather concierge.com/the-great-snow-in-1717-created-great-hardships-in-the -colonies.

Moran, David. "Freezing Rain and Its Effects on Power Lines." DTN, March 28, 2019. https://www.dtn.com/freezing-rain-and-its-effects-on-power -lines.

Morgan, Thomas J. "Hurricane of '38 Wrought Unparalleled Destruction, Death on Unprepared R.I." providencejournal.com. *The Providence Journal*, September 20, 2013. https://www.providencejournal.com /article/20130920/NEWS/309209989.

Moritz, Owen. "How a Blizzard Became a Political Storm for New York City Mayor John Lindsay." nydailynews.com. *New York Daily News*, January 12, 2019. https://www.nydailynews.com/new-york/blizzard-political -storm-mayor-lindsay-article-1.816098.

"National Weather Service's History of the Blizzard of 1978." Boston.com. *The Boston Globe*, February 8, 2008. http://archive.boston.com/news/weather

/articles/2008/02/08/national_weather_services_history_of_the_blizzard _of_1978.

"Natural Disaster Survey Report 78-1: Northeast Blizzard of '78." weather .gov. US Department of Commerce: National Oceanic and Atmospheric Administration, n.d. https://www.weather.gov/media/publications /assessments/Northeast%20Blizzard%20of%201978.pdf.

"Navigation." The Blizzards of 1888—The Blizzards of 1888—National Weather Service Heritage—Virtual Lab. https://vlab.ncep.noaa.gov/web /nws-heritage/-/the-children-s-blizzard.

"Navy Evacuates Warships, Planes; Fleet Units at Norfolk Are Sent to Sea— Aircraft Ordered to Safe Havens." nytimes.com. *The New York Times*, September 11, 1954. https://www.nytimes.com/1954/09/11/archives/navy -evacuates-warships-planes-fleet-units-at-norfolk-are-sent-to.html.

Nemethy, Andrew. "At Vermont State Hospital after Irene, Drama—and Many Questions." vtdigger.org. VTDigger, September 14, 2011. https://vtdigger .org/2011/09/08/at-state-hospital-after-irene-drama-and-many-questions.

"The New England Dark Day, May 19, 1780." newenglandhistoricalsociety .com. New England Historical Society, May 19, 2020. https://www .newenglandhistoricalsociety.com/new-england-dark-day-may-19-1780.

Notchey, James. "National Weather Forecast Office—Taunton, MA." noaa .gov. National Oceanic and Atmospheric Administration, January 1, 2001. https://web.archive.org/web/20130130224655/http:/www.erh.noaa.gov /box/hurricane/hurricaneCarol.shtml.

"NWS Boston—The Great Hurricane of 1938." weather.gov. NOAA's National Weather Service, July 28, 2019. https://www.weather.gov/box/1938 hurricane.

"The Ocean Prediction Center and 'The Perfect Storm.'" ocean.weather.gov. National Oceanic and Atmospheric Administration/National Weather Service, January 2, 2019. https://ocean.weather.gov/perfectstorm/mpc_ps _intro.php.

"The October Snow Blitz: What Made 'Snowtober' So Unusual?" NCAR & UCAR News. National Center for Atmospheric Research, November 16, 2011. https://news.ucar.edu/5758october-snow-blitz-what-made-snow tober-so-unusual.

Old Farmer's Almanac. "The Year without a Summer." almanac.com. *Old Farmer's Almanac*, November 21, 2017. https://www.almanac.com/extra /year-without-summer.

O'Leary, Mary E. "Hurricane Irene Leaves 1 Dead, 1 Missing, 760,000 without Electricity in Connecticut." middletownpress.com. *The Middletown Press*, August 17, 2017. https://www.middletownpress.com/news/article /Hurricane-Irene-leaves-1-dead-1-missing-760-000-11876767.php.

Orloff, Charles. "Tell Us How You Remember the Great Blizzard of '78.'" Blue Hill Observatory and Science Center Climate and Weather Observation Research and Education, February 1, 2018. http://bluehill.org/observatory /2018/02/the-great-blizzard-of-78-remembering-its-fury.

"The Park River." bushnellpark.org. Bushnell Park Foundation. http://www .bushnellpark.org/about-2/history-2/the-park-river.

"Patrick: Could Take Days to Restore Power." CBS Boston. CBS Boston, October 30, 2011. https://boston.cbslocal.com/2011/10/30/state-of -emergency-mass-power-outages-top-650000.

Pealer, Sacha. "Lessons from Irene: Building Resiliency as We Rebuild." vermont.gov. Vermont Agency of Natural Resources, January 4, 2012. https://anr.vermont.gov/sites/anr/files/specialtopics/climate/documents /factsheets/Irene_Facts.pdf.

Peirce, Earl S. "Salvage Programs Following the 1968 Hurricane: Berkeley, Calif., 1968." Edited by Amelia R. Fry. archive.org. University of California Bancroft Library/Berkeley Regional Oral History Office. https://stream/salvageprogram6800peirrich/salvageprogram6800peirrich _djvu.txt.

Perley, Sidney. "Historic Storms of New England." Google Books. Google, 1815. https://books.google.com/books?id=Z2kAAAAAMAAJ&pg =PA187&dq=sidney%2Bperley%2Bgale%2B1815&source=gbs_toc_r&cad =4#v=onepage&q=sidney%20perley%20gale%201815&f=false.

"Population in the Colonial and Continental Periods." census.gov. https:// www2.census.gov/prod2/decennial/documents/00165897ch01.pdf.

"The Portland Gale of 1898, and the Cat That Saved a Life." newengland historicalsociety.com. New England Historical Society, November 27, 2019. https://www.newenglandhistoricalsociety.com/the-portland-gale -of-1898-and-the-cat-that-saved-a-life.

"Portland Gale of 1898." arcgis.com. National Oceanic and Atmospheric Administration/The Preserve America Initiative, July 28, 2017. https:// www.arcgis.com/apps/MapJournal/index.html?appid=ee0d2462b0c94e718 41dfbea8601e3c8.

Preston, Chris, and Lisa McCormack. "Irene Lays Waste to Waterbury Village." vtcng.com. Vermont Community Newspaper Group/Waterbury Record,

August 25, 2016. https://www.vtcng.com/waterbury_record/news/irene
-lays-waste-to-waterbury-village/article_582ce504-d7f5-11e0-9862-
001cc4c03286.html.

"Protecting Vermont Covered Bridges in the Wake of Hurricane Irene."
vermonthomeproperties.com. Vermont Home Lodging Properties, 2021.
http://vermonthomeproperties.com/2011/09/protecting-vermont-covered
-bridges-in-the-wake-of-hurricane-irene.

"The Quarterly Journal of Literature, Science and the Arts." Google Books.
Google. https://books.google.com/books?id=HyxGAAAAcAAJ&pg=PA1
02#v=onepage&q&f=false.

Quinlan, John S. "Ice Storm Climatology in Eastern New York and Western
New England." cstar.cestm.albany.edu. National Weather Service. cstar
.cestm.albany.edu/TV/TV_spring09/Ice%20Storm%20Historical%20
Perspective.pdf.

"Regional Snowfall Index (RSI)." ncdc.noaa.gov. National Oceanic and
Atmospheric Administration. https://www.ncdc.noaa.gov/snow-and-ice
/rsi.

Reilly, Wayne E. "Record Heat, Storm Blasted Bangor a Century Ago." bangor
dailynews.com. *Bangor Daily News*, July 11, 2011. https://bangordailynews
.com/2011/07/10/living/record-heat-storm-blasted-bangor-a-century-ago.

"Remembering the 1978 Blizzard." New England Historical Society,
February 18, 2021. https://www.newenglandhistoricalsociety.com
/15-facts-1978-blizzard.

"Remembering the Great Colonial Hurricane of 1635." newenglandhistorical
society.com. New England Historical Society, August 26, 2020. https://
www.newenglandhistoricalsociety.com/remembering-the-great-colonial
-hurricane-1635.

Rose, Ben, and Katherine Ash. "Irene: Reflections on Weathering the Storm."
floodready.vermont.gov. Division of Emergency Management & Homeland Security/Irene Recovery Office. https://floodready.vermont.gov/sites
/floodready/files/documents/2013-IRO-final-report%20reduced.pdf.

Roth, David. Seventeenth Cenutry Virginia Hurricanes. https://www.wpc.ncep
.noaa.gov/research/roth/va17hur.htm.

Roy, Patricia. "Looking Back: The Ice Storm 2008." telegram.com. *The
Landmark*, December 5, 2018. https://www.telegram.com/news/20181205
/looking-back-ice-storm-2008.

Ruane, Michael E. "Maine Storm Frozen in Memory." *The Washington Post.* WP Company, January 20, 1999. https://www.washingtonpost.com/wp-srv /local/daily/jan99/lessons20.htm.

Samenow, Jason. "NOAA: 2011 Sets Record for Billion Dollar Weather Disasters in the U.S." *The Washington Post.* WP Company, December 7, 2011. https://www.washingtonpost.com/blogs/capital-weather-gang/post /noaa-2011-sets-record-for-billion-dollar-weather-disasters/2011/12/07 /gIQAjD9kcO_blog.html.

Sarnacki, Aislinn. "20 Years Later, Memories of Maine's Ice Storm of '98 Still Fresh." *Bangor Daily News,* January 5, 2018. https://bangordailynews.com /2018/01/05/news/state/frozen-in-time-memories-of-the-ice-storm-of-98.

Semon, Craig S. "10 Years Later, Powerful Worcester County Ice Storm Remains Frozen in Memory." thelandmark.com. *The Landmark,* December 8, 2018. https://www.thelandmark.com/news/20181208/10-years-later -powerful-worcester-county-ice-storm-remains-frozen-in-memory.

Seymour, Tom. "The Great Gale of 1898 and the Sinking of the Steamer *Portland.*" fishermensvoice.com. *Fishermen's Voice,* October 2014. https:// www.fishermensvoice.com/archives/201410TheGreatGaleof1898AndThe SinkingOfTheSteamerPortland.html.

Shea, Jim. "Unaware, Then Overwhelmed: Hurricane of 1938 Still Historic." courant.com. *Hartford Courant,* December 12, 2018. https://www.courant .com/courant-250/moments-in-history/hc-250-hurricane-1938-weather -20140125-story.html.

Slater, Martha. "Rochester Cemetery Commission Deals with the Unthinkable." ourherald.com. *The White River Valley Herald,* October 20, 2011. https://www.ourherald.com/articles/rochester-cemetery-commission -deals-with-the-unthinkable.

"Snowstorm Forces Towns to Move Halloween Trick-Or-Treating." CBS Boston. CBS Boston, October 30, 2011. https://boston.cbslocal.com /2011/10/30/snowstorm-forces-worcester-to-postpone-trick-or-treating -to-nov-3.

Special, Lawrence Fellows. "Coliseum Roof Collapses at Hartford Civic Center." The New York Times. *The New York Times,* January 19, 1978. https://www.nytimes.com/1978/01/19/archives/new-jersey-pages-coliseum -roof-collapses-at-hartford-civic-center.html.

Stanley-Mann, Elizabeth. "The 1927 Flood." uvm.edu. University of Vermont, 2005. https://www.uvm.edu/landscape/learn/Downloads/scrapbooks /1927Flood.pdf.

"Steamship *SS Portland*." lighthouseantiques.net. Kenrick A. Claflin & Son. http://www.lighthouseantiques.net/steamship%20portland%20gale.htm.

Stewardson, Jack. "Remembering the Portland Gale." southcoasttoday.com. *South Coast Today*, January 11, 2011. https://www.southcoasttoday.com /article/19981129/news/311299992.

"Surprising Stories: The Great White Hurricane of 1888." New England Historical Society, March 11, 2021. https://wwwnewenglandhistorical society.com/great-white-hurricane-of-1888.

Takahama, Elise. "This Storm Sounds Bad. The Great Blizzard of March 12, 1888, Was a Lot Worse—The Boston Globe." BostonGlobe.com. *The Boston Globe*, March 12, 2018. https://www.bostonglobe.com/metro /2018/03/12/this-storm-sounds-bad-the-great-blizzard-march-was-far -worse/xTFUSaFDII3AmEwCpJzGyI/story.html.

Thomson, M. T., W. B. Cannon, M. P. Thomas, and G. S. Hayes. "Historical Floods in New England." pubs.usgs.gov. United States Department of the Interior/Geologic Survey, April 20, 1964. https://pubs.usgs.gov/wsp/0798 /report.pdf.

"Top 20 Greatest Snowstorms: Albany, NY (1884–85 to Present)." weather .gov. National Oceanic and Atmospheric Administration (NOAA), March 2021. https://www.weather.gov/images/aly/Climate/Storms.JPG.

Toppan, Andrew. "The Portland Gale." hazegray.org. Haze Gray & Underway. https://www.hazegray.org/features/1898gale.

"U.S. Census Bureau QuickFacts: Boston City, Massachusetts." Census Bureau QuickFacts. https://www.census.gov/quickfacts/bostoncitymassachusetts.

"U.S. Tornado Climatology." ncdc.noaa.gov. National Climatic Data Center. https://www.ncdc.noaa.gov/climate-information/extreme-events/us -tornado-climatology.

"Unamed Hurricanes." ncdc.noaa.gov. National Climatic Data Center/ National Oceanic & Atmospheric Administration, December 11, 2016. https://web.archive.org/web/20161212000841/https:/www.ncdc.noaa.gov /oa/satellite/satelliteseye/hurricanes/unnamed91/unnamed91.html.

Villani, Luke. "November 28: The Portland Gale Leaves Connecticut Buried." Today in Connecticut History. Office of the State Historian/CThumanitie, March 11, 2019. https://todayincthistory.com/2018/11/28/november-28 -the-portland-gale-leaves-connecticut-buried.

"Volcanic Explosivity Index (VEI)." usgs.gov. United States Geological Survey. https://www.usgs.gov/media/images/volcanic-explosivity-index-vei-a -numeric-scale-measures-t.

"Volcano World." Tambora/Volcano World/Oregon State University. http://
    volcano.oregonstate.edu/tambora-0.
Wajda, Shirley T. "Eighteen-Hundred-and-Froze-to-Death: 1816, The Year
    without a Summer: Connecticut History: A CTHumanities Project."
    connecticuthistory.org. Connecticut History/CTHumanities Project,
    August 17, 2020. https://connecticuthistory.org/eighteen-hundred-and
    -froze-to-death-1816-the-year-without-a-summer.
"The Wallingford Tornado of 1878 Unleashes Terror on an Unsuspecting
    Town." newenglandhistoricalsociety.com. New England Historical Society,
    August 19, 2020. https://www.newenglandhistoricalsociety.com/the
    -wallingford-tornado-of-1878-unleashes-terror-on-an-unsuspecting-town.
"'Watching the Fire Bow and Scrape': 150 Years: Bates College." bates.edu.
    Bates College. https://www.bates.edu/150-years/months/october/fire
    -of-1947/watching-the-fire-bow-and-scrape.
"'We Were Turned into Fire Wardens': 150 Years: Bates College." bates.edu.
    Bates College. https://www.bates.edu/150-years/months/october/fire
    -of-1947/fire-wardens.
Weber, Terry. "What Really Happened to the *Andrea Gail*?" gloucestertimes
    .com. *Gloucester Daily Times*, October 29, 2011. https://www.gloucester
    times.com/news/local_news/what-really-happened-to-the-andrea-gail
    /article_948952e0-0c80-5f81-9b18-0cc80be6a629.html.
"WFO Taunton Storm Series Report # 2009-01: Analysis of the December
    11–12, 2008 Destructive Ice Storm across Interior Southern New
    England." weather.gov. National Weather Service Forecast Office Taunton,
    March 2009. https://www.weather.gov/media/box/science/December
    _2008_Ice_Storm.pdf.
"When Maine Burned: Remembering 50 Years Ago." firehouse.com. Firehouse,
    August 1, 1997. https://www.firehouse.com/operations-training/article
    /10544832/when-maine-burned-remembering-50-years-ago.
Wilding, Don. "Shore Lore: Recalling 'The Perfect Storm.'" brewster.wicked
    local.com. *The Cape Codder*, October 26, 2016. https://brewster.wicked
    local.com/entertainmentlife/20161030shore-lore--recalling-the-perfect
    -storm.
"Windsor/Poquonock." neam.or. New England Air Museum, April 8, 2016.
    http://www.neam.org/shell.php?page=history_tornado_pt03.
"Winsted CT 1955 Flood of Mad River, Hurricanes Connie and Diane."
    family-friendly-fun.com. http://www.family-friendly-fun.com/stevengprice
    /Winsted-CT-1955-flood/iindex.htm.

"Winter Storms." Northeast States Emergency Consortium. Department of Homeland Security Federal Emergency Management Agency. http:// nesec.org/winter-storms.

"World War II State Park History." riparks.com. Rhode Island State Parks/ Rhode Island Department of Environmental Management Division of Parks & Recreation, 2012. https://riparks.com/History/HistoryWorld WarII.html.

"The Worst Massachusetts Hurricanes of the 20th Century." Mass.gov. Massachusetts Office of Coastal Zone Management. https://www.mass .gov/service-details/the-worst-massachusetts-hurricanes-of-the-20th -century.

"The Worst Weather Disaster in New England History." newengland.com. New England Today, August 18, 2020. https://newengland.com/today/living /new-england-history/worst-weather-disaster-new-england-history.

"The Wreck of the *Angel Gabriel*, 1635." The Wreck of the *Angel Gabriel*: Colonial Pemaquid: History: Discover History & Explore Nature: State Parks and Public Lands: Maine ACF. Bureau of Parks and Lands. https:// www.maine.gov/dacf/parks/discover_history_explore_nature/history /colonialpemaquid/gabriel.shtml.

"Writers' Forum: Government Weather vs. Private Service." news.google.com. *Meriden Journal*, September 1, 1955. https://news.google.com/newspapers? id=g8NIAAAAIBAJ&sjid=zgENAAAAIBAJ&pg=3503%2C38968&dq= hurricane%2Bdiane&hl=en.

"The Year a State Burned: Maine Fires of 1947 Wipe Out 9 Towns." newenglandhistoricalsociety.com. New England Historical Society, September 20, 2020. https://www.newenglandhistoricalsociety.com/maine -fires-1947-year-state-burned.

"Year without a Summer, 1816." celebrateboston.com. Celebrate Boston. http:// www.celebrateboston.com/disasters/year-without-a-summer.htm.

"The 'Ocean Graveyard' Is Filled with Cape Cod Shipwrecks." CapeCod.com, February 28, 2019. https://www.capecod.com/lifestyle/the-ocean -graveyard-is-filled-with-cape-cod-shipwrecks.

Index

About the Author

Eric P. Fisher is Chief Meteorologist for CBS Boston's WBZ-TV News and anchors weather segments weeknights at 5 p.m., 6 p.m., and 11 p.m., as well as WBZ-TV News at 10 p.m. on TV38 (WSBK-TV). He is also a contributor for CBS News, often found reporting on breaking severe weather across the country.

Born and raised in New England, Eric says there are few places on earth that produce weather like this little corner of the U.S. It offers the challenges of blockbuster snowstorms, hurricanes, tornadoes, heat waves, frigid cold snaps, and dramatic seasonal shifts. Eric vividly remembers Memorial Day of 1995 as a day that helped solidify his path as a meteorologist, when an infamous tornado ripped through Great Barrington in the Berkshires. Glued to the red warnings crawling across the screen and watching the radar, his career in weather was born.

Fisher joined WBZ-TV News from The Weather Channel in Atlanta, where he spent three years as a Meteorologist. He produced and delivered national forecasts and contributed to numerous live reports on extreme weather for The Weather Channel, NBC Nightly News, TODAY, and MSNBC. Some of the most notable events Fisher reported from include the Moore, Oklahoma, tornado in 2013; the 2011 tornado Super Outbreak; Hurricanes Sandy, Irene, and Isaac; and the massive blizzards that essentially shut down New York City in December of 2010 and Boston in February of 2013. Recent times have unfortunately provided no shortage of natural disasters. Previously Fisher worked as the morning meteorologist at WGGB-TV in Springfield, Massachusetts.

A supporter of science and learning, Eric is a member of both the Mount Washington Observatory and the Blue Hill Observatory here in New England. Eric is a graduate of the State University of New York at Albany (SUNY) with a BS degree in Atmospheric Science.